THE CHOICE OF TECHNOLOGY IN DEVELOPING COUNTRIES

Some Cautionary Tales

HARVARD STUDIES IN INTERNATIONAL AFFAIRS
Number 32

The Choice of Technology in Developing Countries

Some Cautionary Tales

By

C. Peter Timmer, John Woodward Thomas, Louis T. Wells, Jr., and *David Morawetz*

Published by the
Center for International Affairs
Harvard University

Library of Congress Catalog Card Number 74-26377
ISBN 87674-027-1
Printed in the United States of America

Harvard University
Center for International Affairs
Executive Committee

Created in 1958, the Center for International Affairs fosters advanced study of basic world problems by scholars from various disciplines and senior officials from many countries. The research of the Center focuses on economic, social, and political development; the management of force in the modern world; the problems and relations of advanced industrial societies; transnational processes and international order; and technology and international affairs.

The Harvard Studies in International Affairs, which are listed at the back of this book, may be ordered from the Publications Office, Center for International Affairs, 6 Divinity Avenue, Cambridge, Mass. 02138, at the prices indicated. Recent books written under the auspices of the Center, listed on the last pages, may be obtained from bookstores or ordered directly from the publishers.

About the Authors

C. Peter Timmer is an agricultural economist who has written extensively on the economics of commodities and their markets. He received his Ph.D. at Harvard in 1970 and has taught economics at Stanford University, advised the Government of Indonesia on agricultural development policy and is engaged in research at the Food Research Institute at Stanford, where he is an Associate Professor.

John Woodward Thomas, a specialist on rural development in the less developed countries, holds a Ph.D. from Harvard (1968). Since 1958 he has worked on problems of economic development and political change in Bangladesh, Ceylon, Ethiopia, Pakistan, and the Philippines. In between overseas assignments he has served at Harvard as Associate Director of the Development Advisory Service (Harvard Institute for International Development) and as a lecturer and research fellow. He is currently directing a multinational study on the employment-creating effects of public works programs.

Louis T. Wells, Jr. is Associate Professor of Business Administration at the Harvard Graduate School of Business Administration, from which he received a D.B.A in 1966. He has advised, among others, the governments of Liberia, Ghana, Indonesia, Malaysia, and Egypt, and has acted as consultant to the governments of Colombia and Peru. The author of articles and reports on multinational enterprises, he is particularly concerned with foreign investment policy and with negotiations between foreign investors and host governments.

David Morawetz received his Ph.D. from the Massachusetts Institute of Technology in 1972. He has worked as a consultant and analyst for the World Bank, the International Monetary Fund, and the Australian Government. In 1970-71 he was a member of the Harvard University Development Advisory Group in Colombia, where he specialized in the analysis of problems of the economic integration of the countries of the Andean Group. He is currently a Lecturer in Economics at Hebrew University in Jerusalem.

Contents

THE CHOICE OF TECHNIQUE IN INDONESIA

C. Peter Timmer

Acknowledgments

A first draft of this paper was presented at the Harvard University Development Advisory Service Conference at Torremolinos, Spain, September 1972. It was prepared while the author was a member of the Harvard Advisory Group/BAPPENAS Research Project (Indonesia) May—September 1972 and was substantially revised while a consultant to the Development Advisory Service. I would like to thank Dr. Saleh Afiff of BAPPENAS, William Janssen and Jim Norris of USAID/Jakarta, Gustav Papanek and Carl Gotsch of the DAS, Walter P. Falcon, now the director of the Food Research Institute, and Peter Warr for helpful comments.

THE PROBLEM

A flurry of writings and research by economists on employment and income distribution reflects a growing professional and political judgment that development policies pursued through the 1950s and 1960s, although fairly successful in terms of overall rates of growth in GNP, have largely failed in their implicit goal of making the bulk of poor people better off. The international corps of development economists has gradually perceived that strategies resulting in growing unemployment and poverty are no longer acceptable to concerned economic decision makers in developing countries. New directions are sought.

The response has been a now familiar check list of the obvious sources of labor displacement in a developing economy:

1. Interest rate too low relative to what a free market would set as its value;
2. Foreign exchange (and hence imported machinery) priced well below its opportunity cost (the price paid by the highest bidder, which becomes binding on all others) to the economy;
3. Artificially high wage levels through premature labor union pressure or government regulation;
4. Lack of technological alternatives that employ labor.

Any of the first three items singly or in combination will be sufficient to bias any investment project toward more capital-using and labor-saving facilities than would otherwise be chosen, with consequently less employment creation per dollar of investment and per establishment. These items, then, represent deviations from free market conditions and can cause less than optimal allocation of productive resources. They are popular with economists as a means of explaining chronic employment problems.

Non-economists—many of them engineers—frequently offer the fourth item as explanation for widespread (and rising) unemployment in the face of substantial investment and growth in industrial output. A number of economists are also persuaded that variation in choice of technique offers little hope of employing more people because of essentially fixed factor proportions (between capital and labor) required by modern (and presumably technically and economically efficient) technology. Income transfers (social security, government services, etc.) and public works are then the only means of avoiding social revolution by the "army of the unemployed."

Indonesia presents a perplexing example of the argument just outlined. On cursory examination the country's pricing policies seem more rational than for almost any other developing country. Although a fairly sizable amount of money is available for medium/long-term investments at only 12% per year, the basic interest rate for secured loans from government banks is between 20% and 30% per year; private banks charge considerably more. In view of the fact that inflation before the rice shortage late in 1972 was no worse than in the United States, these levels are probably close to real rates of return on capital in the Indonesian economy.

Secondly, Indonesia's foreign exchange system is one of the freest of all developed or undeveloped countries, and there is no significant evidence that the rupiah is either over- or undervalued at the moment or will be in the foreseeable future. So foreign exchange is priced at approximately its opportunity cost to the economy.

Lastly, although some International Labour Organization-type labor laws on the books have restricted firing practices and multiple shift operations, they are not completely effective within the modern sector and are of almost no relevance in the much larger traditional sector. The sheer weight of surplus labor has kept unskilled wage rates quite low in both urban and rural areas. This is generally true, although unemployment obviously varies somewhat regionally in such a diverse country as Indonesia. The power of labor unions, which were important politically in the Sukarno era, is very limited.

The prices of capital, foreign commodities, and labor seem to be in order, then. Yet the inability to provide jobs does not seem to be caused by the limited range of technology available. Substantial areas of massive public and private investment clearly offer considerable opportunities for increasing employment: road building, irrigation works, building construction, logging, light manufacturing, and food processing, not to mention agriculture.[1] It seems that the job opportunities existed and that the economic policies and incentives were such that they should have been created and filled.

Yet open and disguised unemployment (underemployment) is regarded by the team of economists managing Indonesia's economy as the major social problem facing the government. The concern is caused by two factors: a) the link between unemployment and dire poverty is very direct in Indonesia (see *9*, p.60), and b) the problem seems to be getting worse, despite the very dramatic recovery of the Indonesian economy in the past five years.[2] Job creation has simply not been taking place to anything like the extent suggested above. Something has gone wrong.

Things have gone wrong, I would argue, in three different ways, none of them visible to the casual observer:

1. Both domestically and foreign-funded government projects are prepared and evaluated in an incorrect manner;
2. The economic policies are not quite as good as they appear at first (although still substantially better than in most other countries);
3. Both Indonesian and foreign technicians are biased in their choice of technique.[3]

All three elements have militated against the choice of labor-intensive techniques in Indonesian production (broadly defined to include the government and service sectors). This paper focuses on the third element, the bias of economic and engineering technicians against labor-intensive techniques, by examining proposals put before the Indonesian government for investment in rice milling facilities. But the other two elements are just as interesting because they reveal the strong interplay among traditional economic models of decision-making, institutional incentives, the strategies of financially and politically potent vested interests, and the structure and preferences of international agencies. It is the positive and negative interaction (feedback loops) among these factors that Falcon and Gotsch identify as critical for understanding "what makes the system go" when forces for change are at work. (*6*)

Falcon and Gotsch note that, with respect to mechanization and the seed-fertilizer revolution, "It is not altogether clear how the blame for the diffusion of inappropriate technology should be apportioned between misguided policies followed by international lending agencies, international and domestic commercial interests, general domestic development policies, etc." (*6*, p. 19). The same can be said for the broader choice of technique in Indonesia. The examination of project preparation and evaluation and economic policies in Indonesia is designed to determine the significance of the Falcon-Gotsch "feedback loops." Are "wrong" decisions simple mistakes due to lack of analysis and information—social mistakes but bureaucratically "right"—or do they stem from self-interest at the expense of the public? Questions such as these are not part of traditional economics, and yet they bear most importantly on that fundamental economic problem of choosing an appropriate productive technique. They are too important to skip over in favor of the more formal economic analysis of the rice milling recommendations.

PREPARATION AND EVALUATION OF GOVERNMENT PROJECTS

When investment projects are prepared on a sector-by-sector basis by sectoral planners who do not have an accurate perception of the national capital availability, a bias toward capital-intensive techniques is almost inevitable. In most economies sectoral planners will be faced with a choice of productive techniques with various capital-labor ratios similar to that shown in Figure 1. The critical relationship between increased gross sectoral output and labor intensity (assuming full employment) is shown in Figure 2.

Assume that a sectoral planner has a relatively fixed labor force ($\bar{L}$) to work with in his sector, a situation typical of many agricultural and related sectors where marginal productivity of labor is less than the average, traditional wage rate. In Figure 2 this sectoral planner is faced with five investment strategies involving five different production techniques (A through E) all producing the same *type* of output. But clearly those techniques with lower labor-capital ratios will produce a greater increase in *gross output* simply because of the greater capital input combined with the fixed amount of labor, $\bar{L}$. The fixed amount of labor is not essential for this to be true, but it simplifies the following argument.

Sectoral growth rate targets are usually thought of in terms of increases in gross output, before a realistic cost for the capital is subtracted in order to arrive at the net output. The sectoral planner is able to maximize the growth rate of gross output for his sector (certainly part of his own objective function[4]) by using the more capital-intensive investment paths. In the absence of accurate national guidelines on capital availability or a realistic price on capital, sectoral planners will choose strategy A or B over D or E.

Choices like those posed in Figure 2 raise the specter of trade-offs between output and employment even though the relationship in Figure 2, AE, reflects full employment of the sectoral labor force $\bar{L}$. It is apparent from Table 1 and Figure 1 that if a sectoral planner becomes committed to an investment strategy such as B under the expectation of receiving 5 units of capital, a rational choice resulting in full employment of $\bar{L}$, unemployment will result if only 2 units of capital are in fact available *and the planner remains committed to strategy B.* Output would increase by 6 units *if* 5 units of capital were available but it would increase by 3.8 units with only 2 units of capital, assuming a shift in technique from B to C. But output will increase by only 2.4 units with the additional 2 units of capital if the B strategy is maintained, and in addition 6 units of labor will remain unemployed. The apparent trade-off between output and employment occurs because of a faulty choice of technique, a choice conditioned at the sectoral level by an inappropriate perception of true capital availability to the economy.

This point of view results from a narrow perspective. The substantial literature on the relationship of sectoral to national economic planning attests to the bias resulting from the myopic vision of sectoral

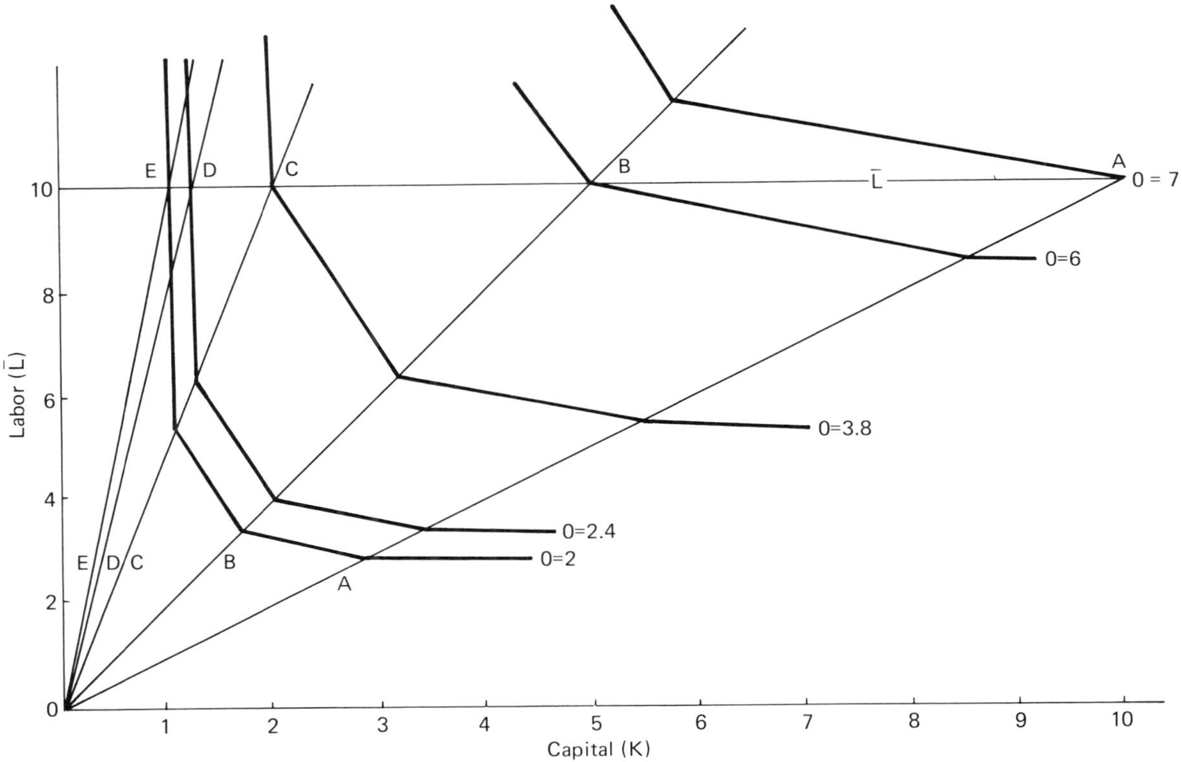

Figure 1—Increase in gross sectoral output according to various productive techniques and alternative levels of capital and labor (from Table 1).

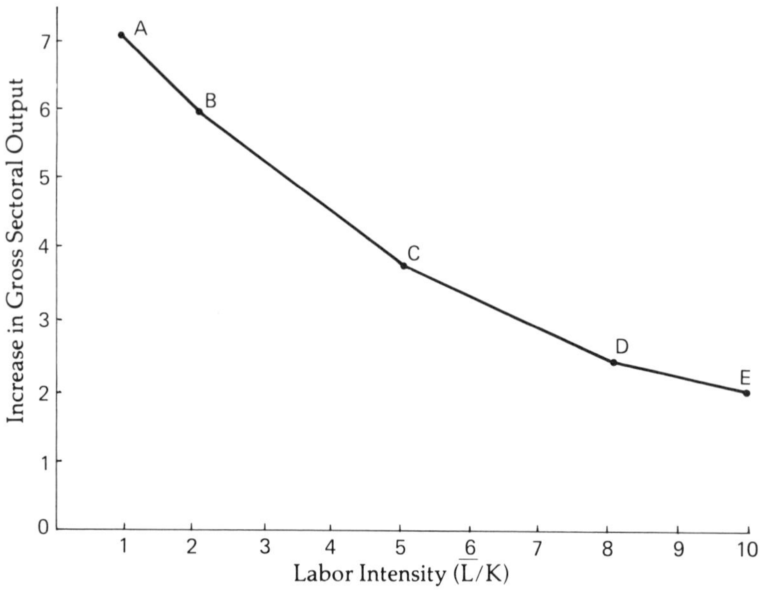

Figure 2—Relationship between increase in gross sectoral output and labor intensity with full employment of sectoral labor $\overline{L}$ and various productive techniques (from Table 1).

planners. Bowles and Falcon, for example, make the following observations:

> In spite of their merits, sectoral formulations also present a series of special problems. The first set of difficulties involves the various linkages to the national models that are necessary to assure consistency of results. These difficulties arise because the economy is a general equilibrium system, and decisions taken in one sector cannot be easily separated from decisions taken elsewhere. Thus, for example, the opportunity costs of inputs used in the sector but supplied from other parts of the economy cannot be known precisely without a full specification of conditions in the entire economy. And these conditions depend, in part, on the decisions taken in the sector under consideration. Similarly, the demand for the output of the sector may depend in important ways on external developments, which may themselves be influenced by choices within the sector. Because of these features, a sectoral model should be thought of as being complementary to, rather than a substitute for, an aggregated model (*3*, pp. 219-220).

If the capital needed to invest fully in technique A (in Figure 2) were available to another sector, then national output would probably be

TABLE 1

Increase in Gross Sectoral Output According to Various Productive Techniques and Alternative Levels of Capital and Labor

Technique	Item	*Increased Output* 2	2.4	3.8	6	7	*L/K*
A	L_A	2.86	3.43	5.43	8.57	10.00	1
	K_A	2.86	3.43	5.43	8.57	10.00	
	$\bar{L}$-L_A	7.14	6.57	4.57	1.43	0.00	
B	L_B	3.33	4.00	6.33	10.00	11.67	2
	K_B	1.67	2.00	3.17	5.00	5.83	
	$\bar{L}$-L_B	6.67	6.00	3.67	0.00	*	
C	L_C	5.26	6.32	10.00	15.79	18.42	5
	K_C	1.05	1.26	2.00	3.16	3.68	
	$\bar{L}$-L_C	4.74	3.68	0.00	*	*	
D	L_D	8.33	10.00	15.83	25.00	29.17	8
	K_D	1.04	1.25	1.98	3.13	3.65	
	$\bar{L}$-L_D	1.67	0.00	*	*	*	
E	L_E	10.00	12.00	19.00	30.00	35.00	10
	K_E	1.00	1.20	1.90	3.00	3.50	
	$\bar{L}$-L_E	0.00	*	*	*	*	

* *Infeasible with surplus sectoral labor force limited to* $\bar{L} = 10$.

increased. Allowing one sector (e.g., modern industry) to invest in a technique similar to A may doom another sector (e.g., agriculture) to a technique like E. Figure 3 shows that from the national point of view technique C maximizes output if everything is measured in market prices, and technique D maximizes output if labor, capital, foreign exchange, and management are appropriately shadow-priced by the government.

This argument is well-known to economists. But its sway over the world is not yet complete. There are countries where guidelines on national priorities, appropriate shadow prices, and even reasonable market prices are not communicated from national planning agencies to the sectoral planning offices. Also, in many countries sectoral planners fail to understand that less than maximum gross output from their sector contributes to maximum national output. Indonesia, in both instances, is such a country.

Accordingly, project preparation and evaluation in these countries, including Indonesia, tend to work in the following fashion. Each sectoral planner determines that the maximum rate of growth in his sector can be achieved by extremely capital-intensive techniques. The Directorate of Fisheries, for example, initially proposed for the second Five Year Plan an investment strategy that would eliminate 100,000 workers from the industry by the end of the Plan (1979), while raising output by more than 10% per year. The strategy involved investing in many large modern fishing trawlers with complete processing and freezing equipment on

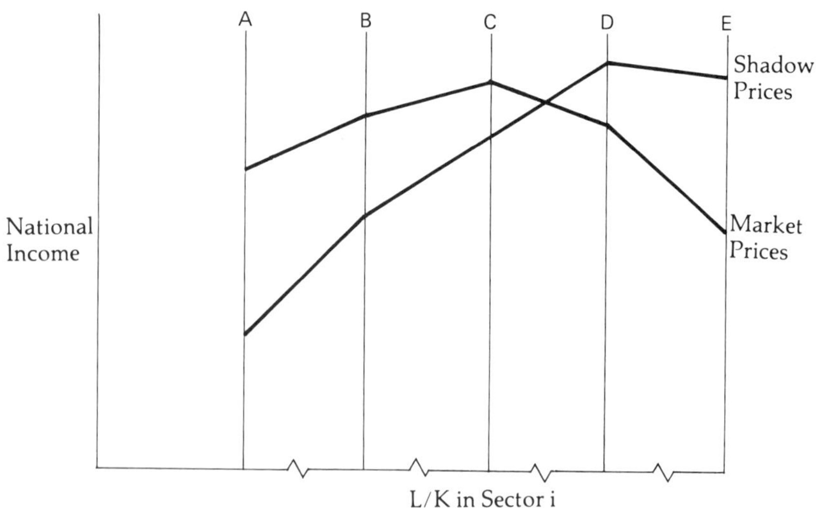

Figure 3—National income according to various investment techniques in a single sector i, assuming appropriate investments in all other sectors.

board. The effect would be to drive the fishermen with sailing proas and small outboard motorboats out of the industry.

When each sector's plans are sent to the center for review, the total investment requirements are some large multiple of the total capital resources available for the development program. Since alternative investment strategies (involving a higher labor/capital ratio) are not presented (and this is obviously the critical failure in the process), the central planning office is unable to make rational choices of technique and capital level by sector. The usual procedure then is either to substantially scale down the program submitted or to send it back to the sectoral planners with a budget ceiling attached.

At this stage, however (and quite possibly all along, for other reasons outlined below), the sectoral planners are wed to their submitted strategy; so they simply scale down the original proposal to meet the budget ceiling. Instead of 100 large fishing vessels capable of raising output 10% per year, only 10 are approved, with output increasing 1% per year. If the resources used to buy 10 large fishing vessels had been used instead for a program of upgrading the traditional fishing fleet and onshore facilities, output could well have gone up by, say, 8% per year.

The process can be clarified by referring back to Figure 2. The sectoral planner may have recognized that technique A was somehow inappropriate for his country at its present stage of development, and so he picked technique B to maximize his sector's gross output. Consequently he submitted a sectoral plan to the central planning office stating that he could increase output by six units if he were allocated five units of capital. Thus, the submitted capital request is five.

The central planning office, when it has received all requests for capital appropriation from the sectors (usually two months late), discovers that the total requests are four times larger than the available funds. Without any clear-cut notion of where marginal productivity of capital is highest the central planning office simply cuts all sector capital appropriations to one-fourth the requested levels. Thus, capital appropriated to this sector is 1.25 although the submitted request was for five units.

It is clear what the sectoral planner should do with the 1.25 units of capital appropriated to his sector. He should drop technique B as an investment strategy and shift to technique D. This permits the sector to reach its maximum output, given its labor force $\bar{L}$ and the newly imposed capital supply. Instead, the planner stays with technique B and simply scales down the investment program to conform to the new capital appropriation. Since only one-fourth of the capital necessary to operate technique B fully is available, it is only operated at one-fourth of "capacity." The resulting increase in output is 1.5 (only one-fourth of the increase possible under technique B if all 5 units of capital had been appropriated). This is less than the potential increase of 2.4 possible with the 1.25 units of capital under technique D.

The labor intensity (L/K) is not 8/1 as it would be under technique D. If it were, there would be full employment (10 units of utilized labor with only 1.25 units of capital). Instead, the realized labor intensity is only 2/1. Only 2.5 units of labor are employed. Three quarters of the total work force remain unemployed, because the relatively capital-intensive technique was chosen.

A major failure in the planning process occurs if alternative investment strategies are not maintained as viable options during the entire evaluation process, and failure is inevitable if alternative strategies are never entertained in the first place. This is a common experience, and the pronounced disinclination of government officials at the sectoral level to search out and consider economically viable labor-intensive projects stems from at least three sources:

1. It is easier from a bureaucrat's point of view to design, construct, administer, regulate, or control a few large capital-intensive projects than many small labor-intensive ones (or even a few large labor-intensive ones).

2. Because the recent hyperinflation made government salaries nearly worthless, corruption of civil servants became an important factor in Indonesian economic life. As elsewhere in the world, government officials in positions of economic power—granting licenses, loans, clearances, controlling customs, etc.—have found it much easier and more remunerative to work with large capital-intensive projects, especially those with a substantial import component, than the smaller, more indigenous, labor-intensive ones. The capacity to generate substantial side payments to regulatory officials is also greater in the capital-intensive industries.

3. Many government investments are financed by foreign aid on the basis of outside feasibility studies conducted by the donor agencies or their contract personnel. The capital-intensive bias this has generated around the world is now well recognized.[5] Indonesia has not been spared. The proposed USAID-financed Bogor highway is a particularly good example. The heavy volume of traffic between Jakarta and Bogor would seem to justify some sort of modern highway. On this basis a USAID contract engineering team recommended a U.S.- style limited access, four-lane superhighway to be constructed strictly with modern engineering equipment and to be maintained in the same fashion. No alternatives were considered or pre-

sented. The Indonesian government (and USAID, which is not entirely pleased with the prospect) is now faced with building either this road or none. This unhappy situation could have been avoided by drawing the attention of the engineers to more labor-intensive alternatives.

The bias against labor-intensive projects financed by indigenous capital resulting from aid-financing stems from several sources. The frequent requirement that the aid funds can be used only for the imported capital component of a project is an obvious one. Less obvious are the greater ease of monitoring large capital-intensive projects by donor agencies, the desire of donor administrators to have a modern, visible monument to show (the Russians especially have this tendency), as well as the requirement that pre-investment surveys and feasibility studies be conducted by Western or non-indigenous (Indonesian-Chinese) technicians. It is this last source of bias, held by technicians, that will be dealt with in the last section of this paper in an examination of a USAID-financed rice marketing study in Indonesia.

ECONOMIC POLICIES AND MARKET SIGNALS

Increasing attention on the part of a number of economic specialists has brought to light serious underlying problems in Indonesian government policies as well as market signals facing entrepreneurs that lead to an anti-employment bias in the choice of technique. The following discussion draws heavily on conversations with, and the reports of, many economists financed by the Harvard Advisory Group, Ford Foundation, FAO, IBRD, USAID, and various foreign governments, as well as with Indonesians within and outside the government.

There seem to be five basic areas of problems. The order of their presentation, although necessarily somewhat arbitrary at the present level of our understanding of the Indonesian economy, corresponds to the author's perception of the relative importance of the problems.

1. The Indonesian tariff structure, if fully administered, would be seriously distorting. The direct distortion is brought about by allowing duty-free imports of capital goods while levying substantial duties on consumer goods and some raw materials and intermediate goods. Firms are thus encouraged to use capital instead of labor, and imported capital equipment rather than domestic equipment. An indirect distortion also exists because duties on some commodities permit a lower price for foreign exchange than would occur otherwise. This hampers exports, promotes imports, and tends to raise the capital intensity of investment.

A more general consideration, possibly the most important, is that high tariffs on goods that are also domestically produced allow the economy to become "inward looking" rather than forcing it to be "outward looking." Empirical evidence is accumulating that the latter strategy generates greater growth in output and employment (and in employment relative to output) than the former. These tariffs may not have had a seriously distorting impact on the present structure of the Indonesian economy because they are alleviated by substantial smuggling. But the tariff structure is distorting current decision-making by foreign and domestic firms.

2. The price and availability of credit is subject to distorting factors. The medium/long-term (up to five years) investment program makes a limited amount of money available at 12% per year, well below other government and private bank rates. One purpose of this program is to promote investment by Indonesian entrepreneurs who are at a significant disadvantage relative to the Indonesian-Chinese entrepreneurs and foreign investors who frequently have access to capital from outside Indonesia (especially Hong Kong and Singapore) at substantially lower rates than the market rates in Indonesia.

Perhaps not unexpectedly, indigenous entrepreneurs have benefited little from the program. The below-market-rate credit must be rationed, and it goes either to very good credit risks (mostly non-indigenous entrepreneurs) or is allocated on the basis of personal or political favor. Apart from who gets the credit, the distorting effects of the subsidized rate of interest can be significant. Rice milling is an interesting example. A number of large modern rice mills (by Indonesian, not world standards) have been built under the medium-term investment program. With a few exceptions they have been built by Chinese or Chinese-backed investors (who were the only investors able to qualify for 12% money). At the same time, a very rapid investment in much smaller facilities—hullers and small rubber-roll rice mills—has been made by indigenous entrepreneurs paying the market rate for their capital, between 2% and 3% per month. The analysis in the last section of this paper will show that these smaller facilities, much more labor-intensive, are far more appropriate for the Indonesian economy than the larger rice mills. Experience to date confirms the analysis. Few of the large mills are able to operate near their planned capacity because the smaller units have outbid them for supplies.

A second major distortion in the credit market is the greater ease in obtaining investment capital relative to working capital. The cheapest working capital available, and then only for firms or individuals able to post 150% of the loan as collateral, is at 24% interest per year. Private rates, e.g., to the milling industry, run about 5% per month. Limits on the availability of working capital seem to be an effective constraint on output for several industries. It follows that raising the amount of working capital available from the banking system for a given loan of investment capital would raise the employment content of any fixed investment significantly.

3. Several aspects of present labor laws and regulations (which are now under extensive government review) tend to convert labor from a variable to a fixed cost and make it easier to use machinery rather than men. Act No. 12 of 1964 prohibits the firing of a single worker without the prior consent of a Regional Committee for the Settlement of Labor Disputes (P4D). If ten or more employees are to be laid off, Central Committee (P4P—in Jakarta) authorization is required.

Act No. 1 of 1951 limits working hours to 7 hours during the day and 6 hours at night. An 8-hour day is granted in most cases upon application to the government, but Act No. 1 discourages multiple-shift operations. Rice mills, for instance, cite this law as the reason for not running night shifts during the harvest season. (An additional factor affecting those mills which use electricity is that the night rate for industrial electricity is almost prohibitive.)

Indonesia has a very protective legislation for female employees: paid leave for the first two days of their menstrual period and up to three months paid leave for childbirth.

These laws and regulations are obviously of no consequence in the traditional sectors. But they do have a significant impact on investments in the modern sector, especially for the highly visible foreign and government firms. It is often said but worth repeating that it is the welfare of the total work force that the laws should protect, not just the privileged segment that receives jobs in protected areas of the economy.

4. Indigenous management talent is at least as scarce as capital in Indonesia. There is no good empirical evidence yet on whether capital-saving projects require less management talent than labor-saving projects. In rice milling the management requirements of the small enterprises that are also capital-saving are so low that ample supplies of indigenous managers have been forthcoming. This would almost certainly not have been the case if much larger, more complex facilities had been built. On the other hand, there are no doubt industries where capital can substitute for management, and for these investment in management training is likely to have a high

payoff by making it possible to substitute managerial expertise for capital.

There is some feeling that Indonesian management training also introduces an anti-employment bias. The focus of this training is process-oriented rather than organization-oriented. The effect is that Indonesian management trainees know how to run factories with machines but not with people. Since the chief bottleneck to highly labor-intensive projects is usually administrative and organizational, this is a serious problem indeed.[6]

5. Last is the problem of red tape. Foreign investors in Indonesia must contend with complex forms and procedures and be prepared to wait for months or years for processing before their project can be started. If the firm and the total size of the investment are large enough, these essentially fixed-cost barriers to entry are worth crossing. But these firms tend to be the most capital-intensive in structure and outlook (and frequently carry overtones of economic imperialism). The very firms that are most desirable—small labor-intensive electronics, clothing, furniture-making concerns primarily interested in export—have little political impact and so cannot afford to overcome the obstacles mentioned above. This systematic exclusion of smaller firms substantially reduces the labor intensity of foreign investment. The same argument applies, in lesser degree, to domestic investment as well.

ENGINEERS, ECONOMISTS, AND CHOICE OF TECHNIQUE IN RICE MILLING

Engineers are normally not economists, nor are economists businessmen. But economists try to create models explaining and predicting the behavior of businessmen, and engineers make economic decisions, implicitly or explicitly, in most aspects of their work. The difficulty arises if these three groups use different optimizing criteria. The Western or Western-trained engineer tries to maximize output per unit of relatively scarce input(s), such as yield per acre or milled rice per unit of rough rice. An implicit goal has been cost minimization, which in Western societies has usually meant minimizing labor inputs. Growth in real GNP has usually been the goal of Western economists. They have advocated a strategy of labor-saving investment on the simplistic theory that this kind of investment will result in higher profits for capital owners, that these owners will have a high savings rate, and that they will reinvest sufficiently heavily to stimulate economic growth. While the conventional wisdom is that the businessman wants to make money, the question whether he primarily wants to maximize profits, rate of return, or growth of the firm, or whether he is satisfied with a comfortable life has not been settled.

The results of different decision criteria among engineers, economists, and market entrepreneurs have been noted by Abbott:

> Many recent recommendations on investment in silo construction for grain storage bear the stamp of engineering concepts of efficiency in the countries where large volumes of grain are handled in bulk. Cost-benefit analysis in terms of the conditions of the developing countries has played a secondary role; where undertaken, the answers have often been cooked, perhaps unconsciously, through the assumptions behind the analysis. In various countries 5,000 to 10,000 ton capacity silos have remained 90 per cent empty for a number of years (*1*, p.384).

That most investment proposals bear the mark of engineering criteria should be no surprise; most such proposals are written by engineers or by economists who come to the engineering analysis at such a late stage as to have a minimal degree of freedom in choosing an alternative technology to those proposed by the engineers. The recent Rice Marketing Study (RMS) conducted under USAID auspices by Weitz-Hettelsater Engineers of Kansas City, Missouri, for Indonesia is a good case study.

In late 1969 a Preliminary Investment Opportunity/Technical document outlining a proposed rice marketing research program was agreed to by both USAID/Jakarta and the Indonesian government. It called for an "investment blueprint" to modernize the rice marketing sector over a 10-year period, to be based on a marketing study similar to ones that had recently been financed by USAID for Brazil, the Philippines, and Bangladesh (then East Pakistan). The company that carried out all three of those studies was also chosen to conduct the Indonesian study.

All three of the earlier studies recommended large concrete bulk storage and processing facilities for rice which have in the meantime been partially or wholly built, in some cases with the same engineering company serving as consulting engineer for the construction. Because the language of the document and earlier company discussions with USAID officials had revealed a potential willingness on the part of USAID to finance similar facilities in Indonesia, the RMS team arrived in the country with a rough picture of the recommendations desired by USAID already drawn.

The Indonesian Steering Committee directing the team's efforts was chaired by a policy-sensitive economist from BAPPENAS, the

Indonesian National Planning Agency. Consequently, the team received strong instructions that any recommended investments in new facilities would have to be justified on the basis of Indonesia's factor availabilities, especially the widespread availability of cheap unskilled labor. Yet in its "Advance Draft" submitted in December 1971, the team recommended a $65 million investment package that contained the items shown in Table 2.

The "Advance Draft" contained extensive engineering and cost data for each of the facilities recommended which was used in the original choice of technique analysis (*9.* pp. 67-81). The four different types of milling/storage/drying facilities vary greatly. The K-1 terminals contain 15,000 tons of bulk storage in vertical steel silos for rough rice and conventional gradual reduction milling machinery with a capacity of 9 tons per hour *gabah* throughput. The continuous flow dryer has a capacity of 20 tons per hour. The H-1 satellites are similar, with only 4,500 tons of storage in vertical steel silos and 3 ton per hour milling and drying capacity.

The G facility utilizes a Japanese-type self-contained rice mill with rubber roll huskers. It has 756 tons of *gabah* storage capacity, of which 456 tons are in upright bulk steel bins. The mechanical drying capacity is 3/4 ton per hour; the milling capacity is 1.0-1.1 tons per hour of *gabah*. A sun-drying pad is also provided. The machinery for the K-1, H-1, and G facilities must be imported.

The flash husker with pearler/polisher is locally manufactured by a number of companies. The husker unit has a *gabah* capacity of 3/4 ton per hour. It is linked to an Engelberg-type polisher with an input capacity of 1/2 ton per hour. A thresher shed and sun-drying pad are provided, but no additional storage capacity.

In the original paper presenting an analysis of choice of techniques, a tandem huller unit (A) was substituted for the flash husker with pearler/polisher (C) because the listed capacity of the latter unit seemed too high while the labor requirements were too low. The C unit had lower capital *and* labor costs than all other milling techniques except hand pounding (Z) on the labor-intensive extreme and the K-1 bulk terminals on the capital-intensive extreme. The substitution of A for C permitted a wider potential set of techniques. Unfortunately, the huller is widely regarded as producing an inferior quality of milled rice and as wasting the by-products (though this is not true when it is well maintained and operated in tandem as in the A setup). It is now possible to reintroduce the more desirable C unit with corrected capacity and labor requirements. This should alleviate concern over poor quality rice and loss of by-products.

The procedure followed is to construct an isoquant in value added

TABLE 2
Recommended Investment Package

Number of Units		Type of Facility	Total Storage MT[a] Gabah[b]	Total Cost Equivalent $US[c]
9	(K-1)	15,000 MT Bulk Terminals	135,000	$22,623,478
59	(H-1)	4,500 MT Bulk Satellites	265,500	26,743,697
122	(G)	Japanese-type Rice Mills	92,232	11,042,342
420	(C)	Flash Husker with Pearler/Polisher	42,000[d]	3,870,000
31		Flat Warehouse	93,000	1,101,306
			627,732	$65,380,823

a MT = metric ton

b *Gabah* is threshed stalk *padi.* In international nomenclature it would be "paddy" or "rough rice."

c Exchange rate then only Rp. 380 = US $1.00

d Assumes a storage capacity of 100 tons in the 16 meter square thresher shed.

(created by subtracting the cost of the rough rice input from the value of the milled rice) in the two dimensions of investment cost and number of unskilled workers required to run each facility. It is, of course, possible to construct the isoquant in gross value of milled rice directly instead of value added. However, an isoquant in gross value of milled rice ignores the fact that capital and labor not only substitute for each other; they also substitute for the rough rice input because of the different technical conversion rates of each facility. The difference between the rough rice input price and the output price for milled rice is not fixed—if it were, either isoquant would give the same result. The difference is a function of the government's price and import/distribution policies and the type of rice milling facilities in use. Competitive pressures, which vary significantly from region to region in Indonesia, are also an important factor in setting the difference between the prices of rough and milled rice.

The only way to examine the impact of smaller or larger differences

—and, as noted, the government has a great deal to say about them, so the issue is of considerable relevance to the policy makers—is to include all three inputs (rough rice, labor, and capital) in the analysis. In the two dimensions of the printed page an isoquant in capital and labor makes the most sense, and the rough rice input is handled indirectly by converting the value of output from gross value of milled rice to value added by milling. Only in this way can some very relevant policy questions be treated within the isoquant framework. Once the isoquant is constructed, labor costs are related to investment cost by calculating the present value of an unskilled worker's wage for his lifetime (50 years)—assuming the life of the plant is the same. This approach is the inverse of the more normal technique in which the discount rate is applied to the capital investment to calculate annual capital charges which can be compared with the annual labor bill. The answers are obviously the same either way, but the approach here has the advantage of keeping the relative investment magnitudes required out in the open. This approach is admittedly biased *against* labor-using techniques because maintenance and operating costs (required for the capital equipment) of each facility are ignored.

Table 3 shows the step-by-step calculations that lead to the capital/labor coordinates of the isoquant. Two points in addition to the extensive footnotes to the table should be noted. First, the more capital-intensive facilities eliminate very substantial physical losses of rice in the drying/storing/milling sequence. It is assumed that stalk *padi* or *gabah* is bought wet from the farmer and dried at the mill by either sun-drying or mechanical equipment. Although there is considerable dispute over the losses due to sun-drying when properly done, these are presumed substantial in this analysis. The *physical yield* of milled rice from 1000 tons of *gabah* in a K-1 unit is assumed to be 17.5% higher than for hand-pounding and 13.6% higher than from the husker/polisher. Second, the more capital-intensive units are assumed to yield a much higher quality of rice than either hand-pounding or the small husker/polisher unit. Thus the K-1 bulk terminal unit is capable of selling its rice for Rp 50 per kg, 25% more than hand-pounded rice sells for and 11% more than rice from the husker/polisher unit brings. This greatly overstates the extent of consumer preference for high quality rice (as opposed to certain varieties, which command substantial premiums) in Indonesia, but these financial losses to labor-intensive techniques are included because they do occur in most developed countries.[7]

The overall effect, then, is to permit the K-1 bulk terminal facility to achieve 46.9% higher total financial returns (see Value of Output on Table 3) from each 1000 tons of *gabah* processed than hand-pounding realizes, and 25.9% higher returns than from the husker/polisher unit.

TABLE 3

Derivation of an Isoquant in Value Added

		Hand-pounding Z(s)[a]	Husker/ Polisher C(s)[a]	Integrated Rice Mill G(s/m)[a]	Bulk Satellite H-1(m)[a]	Bulk Terminal K-1(m)[a]
Per Plant						
	Milling Capacity[b] (Tons per year)	—	1,000[c]	2,500	7,200	21,600
	Investment Cost[d] (U.S. dollars)	0	$8,049	$90,511	$453,283	$2,605,926
	Operative Laborers (number per shift)	—	12[c]	16	27	39
Per 1000 Tons of Gabah Input per Year						
	Investment Cost (U.S. dollars)	0	$8,049	$36,204	$62,956	$120,645
	Operative Laborers[e] (number)	22.00	12.00	6.40		1.81
	Milled Rice Output (tons)	570	590	630	650	670
	Market Price (Rp per kg)	40.00	45.0	48.0	49.5	50.0
	Value of Output (million Rp)	22.8	26.6	30.2	32.2	33.5
	Value Added[f] (million Rp)	4.8	8.6	12.2	14.2	15.5
Per Rp 10 Million in Value Added per Year						
	Investment Cost (U.S. dollars)	0	$9,359	$29,675	$44,335	$77,835
	Operative Laborers (number)	45.83	13.95	5.25	2.64	1.17

[a] The "s" in parentheses indicates the facility uses sun-drying; "m" indicates mechanical drying.

[b] Milling capacity is measured in tons of rough rice (*gabah*) input per year, assuming the facility operates eight hours per day and 300 days per year.

[c] The capacity of the C facility as calculated from RMS data is 1,350 tons per year (the effective capacity limitation is in polishing, not husking, so the *gabah* capacity per se is even higher). The unit is presumed to run at only three-quarter capacity in order to raise the quality of output and provide better separation of by-products. Labor requirements have been raised by 50% because the RMS study underestimated the number of laborers required to sun-dry this volume of *padi* or *gabah*. The cost figure is the same as shown in the study.

[d] Costs of buildings and machinery (including a diesel-powered thresher) are included, but not land.

[e] This assumes that one worker hand-pounds about 150 kg of *gabah* per day with a yield of approximately 95 kg of rice.

[f] Value added is calculated assuming a *gabah* cost of Rp 18 per kg which is above the Rp 16 per kg floor price for "village dry" *gabah* at the mill. The cost (and value added) of reducing the moisture to "mill dry" is included as a mill activity.

These savings in losses (less rice is wasted) are greater than even Western engineers claim for their most sophisticated equipment. And yet, when Indonesia's relative factor costs are related to the isoquant constructed from the data in Table 3, as shown in Figure 4, the result is the economic dominance of the labor-intensive husker/polisher.[8]

Although Figure 4 is in many ways the end of the story, it is interesting to pursue the issue a little further. Table 4 calculates the present value of labor costs at alternative wage rates (of which the highest, two and a-half times the market wage for unskilled labor, is used in Figure 4) and the maximum wage at which each milling technique is optimal as *gabah* prices are increased. If the retail price is held fixed, raising the *gabah* price will progressively drive out of operation the more labor-intensive facilities due to their assumed high physical and monetary losses.

Several features are striking in Table 4. The effect of discounting (not a favorite technique of engineers) to obtain the present value of wages is dramatic. The $200 wage for 50 years is a total wage bill of $10,000, but its present value is only $1661 when discounted at 12% per year, the subsidized rate of interest for medium/long term investment credits in Indonesia. Using a market interest rate reduces the present value wage to $833. Market wages discounted at market interest rates have a present value of $333, and a shadow wage rate equal to half the market wage (close to a subsistence income for an individual in rural Java) has a present value of *less than $200.*

Next, at either prevailing market wages or even the high wage assumed by the RMS team for its cost/benefit calculations, the husker/polisher C unit is economically preferable to all other alternatives for all reasonable *gabah* prices. The integrated rice mill (G) becomes optimal when *gabah* prices reach Rp 22 per kg, money is available at only 12% per year, and the investor is paying $200 per year wages for unskilled labor. Thus the economic analysis performed here corroborates the evidence from the countryside: small husker/polisher type units are installed if market prices for inputs and output prevail, but the larger integrated rice mills become (barely) optimal if these prices are sufficiently distorted, especially by subsidized credit. In deficit areas supplied with imported rice and a well-developed private rice trade, differences between *gabah* and retail rice prices may be so small that only the more capital-intensive units are technically efficient enough to survive. But they do so at very low profits and at the pleasure of the government's rice price policy (described in *2*).

Lastly, the economics of hand-pounding rice in Indonesia are fascinating. If market wages must be paid for hand-pounding, it is not profitable unless the discount rate is 27% for *gabah* costing Rp 18 per kg,

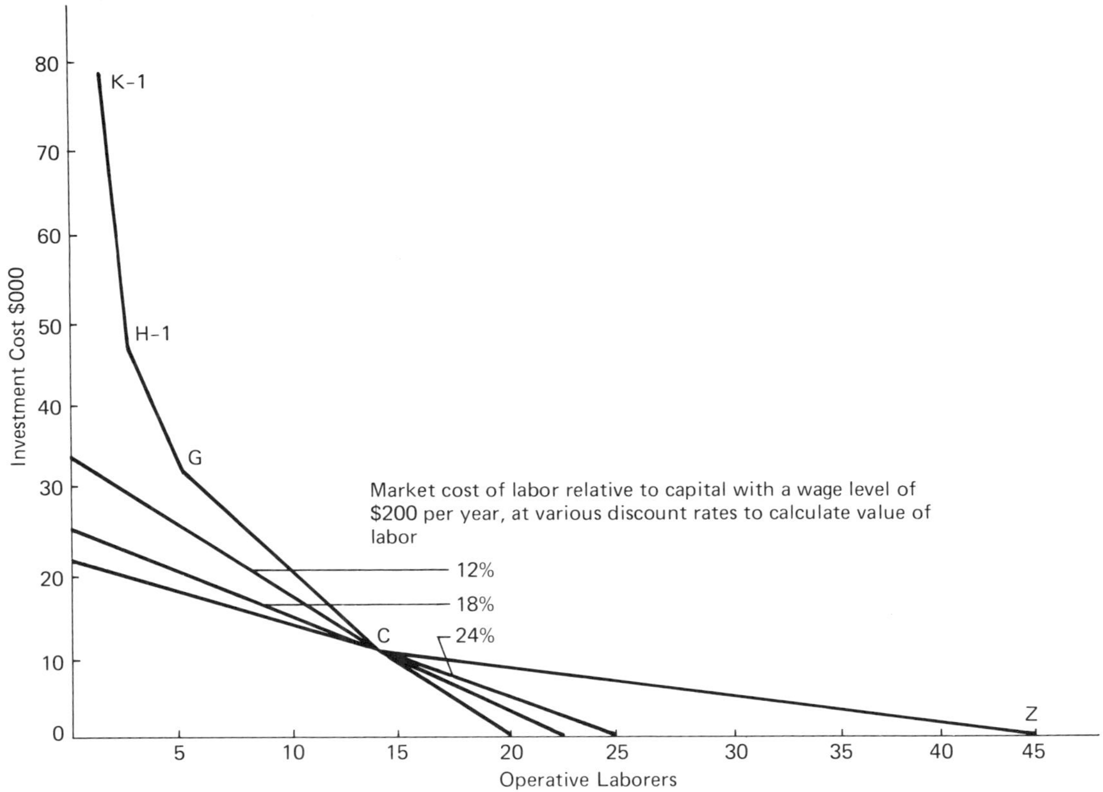

Figure 4—Isoquant relating investment cost and operative workers needed to produce Rp 10.0 million in value added assuming *gabah* costs Rp 18 per kg.

TABLE 4
Present Value of Alternative Wage Rates and Maximum Values that Yield Corner Tangencies (U.S. Dollars)

	Discounted Present Value of Wages [a]		
Assumed wage levels per year	12 percent discount rate	18 percent discount rate	24 percent discount rate
$200 (Rice Marketing Study)	$1,661	$1,111	$833
$80 (Market)	664	444	333
$40 (Shadow Price)	332	222	167

	Maximum present value wage (in dollars) at which the indicated facility is optimal for various *gabah* prices					
Gabah prices (Rp per kg)	18	20	22	24	26	28
Hand-pounding (Z)	294	202	70	b	b	b
Husker/Polisher (C)	2,335	1,956	1,457	766	-260	b
Integrated Rice Mill (G)	5,617	5,034	4,265	3,197	1,669	-728
Bulk Satellite (H-1)	22,789	21,829	20,468	18,879	16,296	12,316
Bulk terminal (K-1) [c]	. . . . greater than the value shown for H-1					

[a] The present worth factor for 50 years is as follows:

Discount Rate [%]	*Present Worth Factor*
12	8.304
18	5.554
24	4.167

[b] Negative value added.

[c] Just for comparison between Indonesia and the United States, it is worth noting that the discounted present value of $6000 per year at 8 percent is $73,400. Thus the relative capital intensity of even the large bulk terminals (K-1) is low by United States standards. Obviously, the RMS did allow Indonesia's relative factor availabilities to influence their recommendations, but not enough.

40% for Rp 20 per kg *gabah*, and 114% (which is nearly 10% per month without compounding) if *gabah* prices are Rp 22 per kg. It is not hard to see why hand-pounding as a cash hire activity is rapidly dying out, even in the poorer rural regions of Java. And yet it is still frequently profitable for the farmer (or more likely his family) to do it himself if the perceived opportunity cost of the labor is not great. For example, hand-pounding is economically optimal (with all its assumed losses) if the wage rate is $40 per year and is discounted at 18% or higher, at least for *gabah* prices less than Rp 20 per kg (as they are likely to be in the rural areas). And, in fact, many rural families still hand-pound their own rice.

In light of this analysis, consider the set of recommendations made in the "Advance Draft" (see Table 2). Three-quarters of the proposed $65.4 million investment is for K-1 and H-1 bulk facilities. The latter is the optimal technique only under artificially high *gabah* prices, and the K-1 facilities never become optimal in Table 4. A further one-sixth of the investment is for integrated rice mills (G) that need subsidized credit or special market conditions to be viable although they clearly fit the Indonesian context better than the bulk facilities. And less than one-twelfth of the funds are allocated to what must be regarded as facilities appropriate for Indonesia's economic environment: husker/polisher mills and flat warehouses.

Despite some efforts to adapt the recommended facilities to Indonesian circumstances (e.g., steel silos rather than concrete), the RMS team failed in the "Advance Draft" to fulfill the Steering Committee's mandate. The reasons were many and complex, involving historical accident and personalities, but beneath all these lay a deep-felt bias on the part of Western and Western-trained technicians that identified capital-intensive with modern, and modern with good. Consider that the RMS team recognized the unprofitability of its bulk facilities—its own benefit-cost ratio for the K-1 facilities, assuming a 12% discount rate (the highest used in the study) and 7% reduction in losses, was *0.41*! In other words, the cost was twice the benefit. In the face of this, the team recommended the K-1 and H-1 facilities because they were necessary to *modernize* (the team's word) Indonesian rice marketing. Bias cannot run much deeper than that.

The Steering Committee was able to convince the team that it should present engineering and cost data on a very wide range of milling, storing, and drying facilities. The provision of such data was a substantial improvement over earlier studies. In addition, many of the economic calculations were performed as requested. Yet, basically, the "Advance Draft" failed because the team ignored its own economic conclusions.

The reasons why it did so are important for an understanding of how socio-political-economic systems function. The engineering company wanted to recommend facilities that it might have some role in building. But beyond this easily understood (and easily discounted) motive was a genuine belief that the Indonesian rice marketing sector would never amount to a hill of beans unless it started some unprofitable *modernization*. It is difficult to be altogether objective about this approach to investment planning. Fundamentally different value judgments are brought to bear in the discussion, and no rational answer emerges. In the end it must be the Indonesians' values that prevail—it is their country. And so it is with the Rice Marketing Study. The "Advance Draft" was rejected by the Steering Committee, essentially for the reasons outlined above, and the team was told to put more confidence in its economics. What has emerged is a final report that backs off considerably from the "Advance Draft." A very wide range of different types of facilities are offered, with costs and benefits neutrally reported, and an outline of the incentive package deemed necessary to interest the private trade in each facility. The incentives required for the big bulk facilities are substantial, and there is no recommendation that they be offered. But if the Indonesian government wants to go down that road, it will have some idea of the costs.

Perhaps the most remarkable aspect of the whole episode is that the ineluctable forces that lead to investment in inappropriate technologies are not so ineluctable after all. The policy makers must be concerned about the issue, they must know about the pressures and biases, and they must do their homework. Given the hectic pace of any good planning agency, this is a lot to ask. But it is not impossible. Despite the importance of the Falcon-Gotsch "feedback loops," they are influential but not determining.

CONCLUSIONS

Indonesia is much better off than it might be with respect to biases in the choice of technique. Factor prices relative to availabilities are good by international standards. Major opportunities exist for profitable employment creation. Indonesia's shortcoming is relative only to the magnitude of the task: 10 million unskilled jobs could probably be filled quickly if they were available.

In the face of this kind of problem any source of bias against the use of labor must be regarded as serious. And all of the above discussion shows that the cards are indeed stacked against labor-intensive, capital-saving technologies, despite the relatively advanced policies of the government.

It is important, then, that concerned governments and policy makers take whatever positive steps are available to counteract this long-standing bias. These include:

1. Making labor-intensive alternatives mandatory in the terms of reference for investment surveys;
2. Carefully analyzing macro-economic policies from an employment maximization point of view with the intent of making positive recommendations consistent with the government's administrative capacities and ideologies;

3. Providing the entire government planning hierarchy with economically rational guidelines for project preparation and evaluation.

Perhaps these measures seem trivial relative to the immense problem at hand. But as political reactions to rising unemployment force government policy makers into action, it is important that the actions strike at the cause and not at the symptoms. Firm implementation of the above three measures would do this, and would ensure no further deterioration in the unemployment situation. Coupled with active programs of job creation such as the Kabupaten rural works projects, these measures would provide a wide-ranging employment-oriented government policy that would promise real hope for Java's underemployed laborers.

Notes

1. The one major exception to this is the massive investment in the petroleum sector. It offers little scope for the use of unskilled labor in any of its operations until creation of the infrastructure after wells have come into production.

2. Only rough estimates are available but they suggest that real per capita incomes have been growing at about 5% per year since 1969. Population is increasing approximately 2.5% per year, although the number of new entrants to the labor force is growing somewhat faster.

3. This is not due to foreign technicians basing recommendations on their own inappropriate backgrounds. The issue here is that even when exposed to the actual conditions in Indonesia and told to make recommendations appropriate to relative factor availabilities, the technicians make recommendations that are *still* inappropriate because of a deep-seated bias against labor-intensive techniques.

4. "Requiring resources used on government account to meet the productivity test of nongovernment user valuation is likely to be a useful rule. Officials may euphemistically be called 'public servants,' but it would be gratuitous to assume that their actions are aimed solely at the maximization of a social welfare function. *Officials have their own welfare functions* and if resource prices are left to their discretion rather than being given to them as objective data, they may set resource values to maximize their welfare function." (*10*, p. 488.) Emphasis added.

5. For agricultural and related technology the following references are indicative:

"Thus, except for a small number of imports that may have come in under cash arrangements in 1965/66 and those imported under barter arrangements, the 15,000 tractors (in West Pakistan) purchased since 1965 (approximately 75 per cent of the total in service at the end of 1971) were obtained with foreign aid loans. Of the total machinery loans, IDA contributed roughly 65 per cent, the Western Europeans, 25 per cent and the USSR, 10 per cent.

"As the concluding sections on mechanization policy will emphasize, the future availability of such funds, earmarked for a specific purpose and not subject to the full scrutiny of the resource allocation process, are likely to determine in large measure the appropriateness of the country's mechanization program." (*4*, p. 13.) A rather dim view of the World Bank's supplying tractors to West Pakistan is contained in 7, which was done by an IBRD staff member and consultant.

"Yet another factor in the unemployment picture is what might be termed a capital bias on the part of aid givers, foreign investors, and many public agencies in less developed countries . . . Preferences for big projects, large machinery, and centralized control, appear to be the rule rather than the exception. The concentration of the aid-giving agencies on the rate of return, the reluctance to shadow-price labor at below market prices at the choice-of-techniques stage in feasibility studies, the preference for foreign contractors, and the inattention to the number of farmers served and the distribution of benefits among those who are involved, have all had their employment costs." (*5*, pp. 13-14.)

"It should be candidly admitted that donor concentration on projects, specifically the foreign-exchange component of projects, often works in just the opposite direction, i.e., induces import and capital intensity . . . .

"In the agricultural sector, aid agencies like The Bank (IBRD) should lean against subsidies for tractors and other heavy equipment in the guise of low duties, and/or low excises on domestic production—in addition to the overvalued exchange rates which favor their importation in the first place—threatening uneconomic and wholesale mechanization in the wake of the Green Revolution." (*8*, pp. 22-25.)

6. It is hoped that the Ford Foundation's management training program will provide a clearer picture of the extent of this problem in the near future.

7. The interested reader is referred to my earlier article (*9*) for a line-by-line discussion on how the isoquant is constructed.

8. The biases against labor-intensive facilities were introduced into the analysis to persuade the skeptical by conceding in advance any possible objections. More realistic assumptions made hand-pounding somewhat more attractive, but not enough to become optimal at market prices.

References

1. J. C. Abbott, "The Development of Marketing Institutions," in *Agricultural Development and Economic Growth*, ed. by N. M. Southworth and B. F. Johnston (Ithaca, New York, 1967).

2. Saleh Afiff and C. Peter Timmer, "Rice Policy in Indonesia," *Food Research Institute Studies*, X, 2, 1971.

3. Samuel Bowles and Walter P. Falcon, "Introduction to 'Sector Planning,' " in *Studies in Development Planning*, ed. by Hollis B. Chenery, et. al. (Cambridge, Massachusetts, 1971).

4. Carl H. Gotsch, "Tractor Mechanization and Rural Development in Pakistan," Harvard University, June 1972.

5. Walter P. Falcon, "Agricultural Employment in Less Developed Countries: General Situation, Research Approaches and Policy Palliatives," IBRD-IDA Economic Staff (Consultant) Working Paper No. 113, April 1971.

6. Walter P. Falcon and Carl H. Gotsch, "Technology, Institutions and Rural Development: Notes for Discussion," presented at the Harvard University Development Advisory Service Conference, Torremolinos, Spain, September 1972.

7. John P. McInerney and Graham F. Donaldson, *The Consequences of Farm Tractors in Pakistan* (An Evaluation of IDA Credits for Financing the Mechanization of Farms in Pakistan, 1966-1970), IBRD-Development Economics Department, January 1973.

8. Gustav Ranis, "Employment and Income Distribution in Developing Countries," IBRD-IDA Economic Staff (Consultant) Working Paper No. 115, September 1971.

9. C. Peter Timmer, "Employment Aspects of Investment in Rice Marketing in Indonesia," *Food Research Institute Studies*, XI, 1, 1972.

10. Richard S. Weckstein, "Shadow Prices and Project Evaluation in Less-developed Countries," *Economic Development and Cultural Change*, May 1972.

THE CHOICE OF TECHNOLOGY FOR IRRIGATION TUBEWELLS IN EAST PAKISTAN: AN ANALYSIS OF A DEVELOPMENT POLICY DECISION

By John Woodward Thomas

Acknowledgments

The present essay has been substantially improved by the comments and suggestions of Stephen V. Allison, Eric Griffel, Walter P. Falcon, Carl Gotsch, Millard Long, Edwin Smith, Joseph J. Stern, and Peter Warr. Their contributions are gratefully acknowledged.

Glossary

ADC	The Agricultural Development Corporation
Amon	The rice growing season from July to November
Aus	The rice growing season from May to July
Boro	The irrigated rice season from December to May
Cusec	The engineering abbreviation for "cubic feet of water per second"
IBRD	International Bank for Reconstruction and Development (World Bank)
IDA	The International Development Association, the concessionary lending arm of the World Bank
KTCCA	Kotwali Thana Central Cooperative Association
Maund	A unit of measure equal to 82.7 pounds
Rupee	The currency of East Pakistan until 1971, equal to $0.21
Thana	Unit of local government roughly equivalent to a county
WAPDA	The Water and Power Development Authority

Development analysts have often suggested that the developing countries share one advantage which the high-income nations did not enjoy: they have a whole range of technologies from simple to complex, from low- to high-cost, from labor-intensive to capital-intensive, which allows them to select those best suited to their needs and objectives. In practice, however, this implied freedom of choice may be seriously limited by several factors. This study explores in detail a particular develop-

---

In this paper the name East Pakistan is used because the events analyzed in this essay took place before the outbreak of the civil war in 1971 and the formal creation of Bangladesh.

Much of the technical analysis of tubewell alternatives contained in this paper is drawn from an earlier paper by the author, *The Development of Tubewell Irrigation in Bangladesh: An Analysis of Alternatives.* Any reader wishing to learn more about the technical alternatives discussed in this article is referred to that paper. I am grateful to US AID for financial support for that paper.

ment policy decision, the choice of technology for irrigation tubewells in East Pakistan. It examines the technological alternatives available to the government, evaluates the choice of technology in terms of national development priorities and feasibility of implementation, and analyzes the factors which determined the final choice.

Both developing countries and international aid-giving agencies, while choosing projects and technology partly on the basis of development goals and administrative feasibility also have other implicit objectives which influence their decisions. This study attempts to identify these and their relative influence on decisions and to suggest that a broader frame of reference is needed if technological choices in developing countries are to be understood.

I. THE ROLE OF AGRICULTURE AND IRRIGATION

With a population of 74 million, East Pakistan in 1970 contained more than 1,350 people per square mile, making it one of the most densely populated areas of the world. Over 90 percent lived in the rural areas and most of these derived their income from agriculture. With no significant resources other than land and people, East Pakistan, comprised primarily of the deltaic plain of the Ganges and Brahmaputra rivers, clearly was an agricultural society. It was to this sector of the economy that it was necessary, in the short run at least, to look for economic growth.

The primacy of agriculture in the economy is underlined by the role it played in the 1960s. In 1969, the province had to import nearly two million tons of food grains to support its population.[1] Agricultural labor comprised 75% of the total work force. Although difficult to measure satisfactorily, unemployment was pervasive; the Second Five Year Plan estimated it as 33% of total available man-days.[2] Fifty-seven percent[3] of gross domestic product and 95 percent[4] of export earnings were derived from agriculture.

In the 1960s the Green Revolution first began to affect East Pakistan. Fertilizer and pesticide were imported, seed research was introduced and given high priority, and irrigation efforts were stepped up. Still, the average growth rate of major crop production from 1965 to 1970 was only 2.2%[5]. Because of the great fluctuations in climate and repeated flooding during *amon* (the primary growing season), when 60% of the rice crop was produced, it was difficult to find a technology that would raise output in that season. Hence, attention turned increasingly to irrigated agriculture in the dry season.

Development of irrigation would permit substantial increase of agricultural production in East Pakistan. Only with irrigation is a *boro* (December-May) crop in the six-month dry season possible. Moreover, recent estimates indicate that between ten and fifteen million acres or 50-75% of total crop land may be suitable for irrigated cultivation.[6] With its continual sunshine and the absence of the recurring floods and storms that damage crops in other seasons, the December-May season is potentially the most productive one of the year.[7] In addition to making this third crop possible, facilities for supplemental irrigation allow marginal shifts in the planting and harvest times of the two other annual rice crops, thereby reducing the risk of crop loss in the peak flood season in August and the loss of productivity because of insufficient moisture at the end of the *amon* season in November.

Since 1954 there had been attempts to increase agricultural production through irrigation. These could be classified in three categories. First, there had been a series of large multipurpose projects consisting of flood control, drainage, and irrigation. These were generally carried out by the Water and Power Development Authority (WAPDA), but had yielded only a .008 percent increase in total crop area.[8] Many of them were completed in the late 1960s. Although new projects of this type were planned, few benefits from the larger projects were anticipated until the late 1970s because of their long gestation periods.

Second, surface waters (rivers, canals, etc.) obtained by low-lift pumping increased the amount of irrigated land rapidly in the 1960s: by the end of the decade 27,000 pumps were in use. However, studies showed that available surface water was sufficient for no more than 35,000 pumps which would irrigate only 20% of the total irrigable land. As a result the government of East Pakistan and its aid donors began to give high priority to the third alternative, tubewell irrigation.

Accordingly, the Pakistan Fourth Plan (1970-75) established a target of 20,000 tubewells to be installed in East Pakistan. It budgeted a generous Rs. 1,400 million ($295 million), the equivalent of Rs. 70,000 ($14,750) per well.[9] To meet this target, it was expected that more than half the funds would have to come from external sources and that therefore aid donors as well as the government of East Pakistan would play a role in the selection of the type of well and the method of installation. The choice of technology would have significant social as well as economic consequences because it would determine the number of wells that could be financed, their location, and the way in which the benefits would be distributed throughout the country.

II. PILOT PROJECTS, OPTIONS, AND PLANS

To put the decision on tubewell technology into proper perspective it is necessary to know something of East Pakistan's firsthand experience with tubewells and the range of technical alternatives available.

A. Pilot Projects

At the time the initial choices of tubewell technology were made, the pilot tubewell projects in Thakurgaon and Comilla provided the only firsthand experience with tubewell irrigation in East Pakistan. The two also illustrate the range of choice available. In Thakurgaon, a high-cost tubewell field was constructed and operated under the auspices of WAPDA. The feasibility study, design (including the choice of well location), and construction were done by West German engineers under a West German supplier's credit. The project was located in an area known to have an excellent ground water supply. Three hundred eighty wells of three-cusec (cubic feet per second) capacity were installed and an electric generating plant was constructed solely to provide power for the project. The wells, which were sunk by power drilling rigs, used steel screens and turbine pumps, all from West Germany. The total installation, including the generating plant, took five years, the well field becoming operational in 1965.

The cost of these wells was very high, approximately Rs. 260,000 ($54,750) per well or Rs. 87,000 ($18,250) per cusec of water. Just over half the cost was for electric generation and transmission facilities.

Two major problems arose in Thakurgaon: wells were located on high ground along a ridge and on porous soil that proved inefficient for irrigation, and farmers were not taught to utilize irrigation water. Largely for these reasons, four years after completion of the project only 31% of the irrigable land was being cultivated in the dry season.[10]

In Comilla, wells were sunk by the Kotwali Thana Central Cooperative Association (KTCCA), which adapted for local conditions two low-cost labor-intensive techniques for installing wells: percussion and jet drilling. Two hundred and eleven diesel-powered wells, designed for two-cusec output and using locally fabricated brass screens were installed between 1962 and 1969. The cost of the wells, Rs. 23,700 ($5,000) per well or Rs. 11,850 ($2,500) per cusec, was borne by KTCCA and the Agricultural Development Corporation. Although water utilization has been inefficient, resulting in less than optimal coverage, the wells have proven highly profitable from both a private and a national economic perspective.[11]

Despite the profitability of irrigation in Comilla there has since been little private investment in tubewells. A study of land and rural credit practices in West Bengal (India) shows that moneylenders are not willing to lend money for tubewells since the profitability of wells would usually break the farmers' dependence on the moneylender. There is every reason to believe that this situation existed in adjoining East Pakistan as well.[12] The general level of rural poverty has meant that only very few farmers had an investable surplus, and for a number of reasons even these few well-to-do farmers did not invest in tubewells. For one, government subsidy policies discouraged private tubewell development, even where the value of water was understood. As we have seen, with the help of subsidies KTCCA installed wells for Rs. 23,700 each; based on market costs in Dacca in 1970, a similar well would have cost Rs. 32,000. Actually, the government did not expect farmers to meet any of the capital costs of installation. In Comilla, farmers had only to pay some of the operating and maintenance costs; in Thakurgaon, water was provided free. Thus, in view of the government's widely announced plans to undertake a major new well installation program, even farmers who had an investable surplus lacked the incentive to invest in wells. Moreover, without the help of agencies capable of installing wells the difficulties for pioneers in sinking wells were prodigious. With technical supplies located almost exclusively in Dacca, trained drillers only in Comilla, and without maintenance or service facilities, the problems of attempting to install private wells are obvious.

B. *Options*

There was a wide range of options for well installation and operation in East Pakistan. Comparisons of well specifications can be made by using cost per unit of output (measured in cusecs), since water quality everywhere is similar, and only quantity varies. The factors affecting cost fall into five general categories: (1) drilling technique, (2) power source and type of engine, (3) type of pump, (4) screen material used, and (5) agency installing wells. The major alternatives in each of these categories and their cost are shown in Table 1.[13]

If the major alternatives within these categories and the possible combinations in which they can be used are considered, there are more than 250 possible well designs. But in practice they are used in packages that are homogeneous, in terms both of technology and cost. For example, a well drilled with a simple, low-cost, labor-powered percussion rig will probably not use the most expensive imported fiberglass screens. For the purposes of this paper, wells are grouped in similar cost and technological groups: high-cost wells (over Rs. 100,000) using capital-intensive imported technologies, imported equipment, foreign contractors to install wells, and electric generators to power them; medium-cost wells (Rs. 100,000 to 50,000), utilizing capital-intensive, complex imported technologies and supplies but relying on diesel power, with installation by foreign or domestic contractors; and low-cost wells (under Rs. 50,000), utilizing labor-intensive methods, adapted technologies, and supplies generally manufactured or assembled within the country and in-

TABLE 1

Costs of Major Variable Items in Tubewells
(except cost of installation)
(in Rupees)

	Units	Costs	Foreign Exchange Component	Accounting Price	Source
Drilling	per foot (including depreciation)				
Water Jet		8.00	25%	10.00	KTCCA cost 1970
Manual percussion		10.00	5%	10.50	KTCCA/ADC cost 1970
Power (by contractor)		47.50	75%	83.00	IBRD estimate 1970
Power Source	20 h.p. engine				
Low speed diesel		4,500.00	50%	6,750.00	Author's estimate (assumes domestic production)
High speed diesel		6,000.00	100%	12,000.00	IBRD estimate 1970
Electric		5,000.00	100%	10,000.00	Dacca market price 1970
Pumps	per pump (see Table 2 for more complete comparison)				
Centrifugal		750.00	50%	1,125.00	Dacca market price 1970
Turbine (inc. right angle gear)		7,000.00	100%	14,000.00	IBRD estimate 1970
Screen	8″ diam. per foot				
Poly Vinyl Chloride (PVC)		34.50	10%	38.00	Prices based on IBRD estimates 1971. Foreign exchange component assumes local production.
Fiberglass		125.00	80%	250.00	
Brass		50.00	70%	85.00	Dacca market price 1970

Accounting Price—where market prices do not reflect the real costs to the national economy, accounting prices are used to estimate the true cost of the goods or services.

stalled by domestic contractors, agencies of the government, or agricultural cooperatives.

C. Plans

In 1969-70, East Pakistan was engaged in a debate over the appropriate technology for tubewells. The 1970s were expected to be a time of rapid installation of wells, and proponents of all three categories of wells were arguing their case. Three agencies of the East Pakistan government were involved.

The East Pakistan Planning Department bore formal responsibility for approving all development projects and policies, but it played a minor role in the choice of tubewell technology. After approving plans and financing for the rapid development of 20,000 tubewells the department, which was short of technical and engineering talent, left the crucial technical decisions and the negotiations with aid donors to the operating agencies.

The two operating agencies of the government of East Pakistan vitally concerned with tubewell development were the Agricultural Development Authority (ADC) and the Water and Power Development Authority (WAPDA), powerful semi-autonomous government agencies competing for leadership in this new area of development. WAPDA, founded in 1958 for planning and implementation of the province's water development programs, had been a conduit for an estimated 19% of total development resources, and 31% of foreign aid in the 1960s.[14] Enjoying a near-monopoly of engineering talent, close relations with aid donors, and considerable power in internal government affairs, WAPDA, which was also bolstered by a large corps of foreign advisers as well as experience in large-scale projects, appeared the logical agency to sponsor tubewell development. After implementing the Thakurgaon project, WAPDA had prepared feasibility studies and submitted plans to aid donors for several other high-cost tubewell projects.

Smaller and comparatively less powerful was the ADC. Formed in 1961 as the agency providing agricultural supplies such as fertilizer, pesticides, and new seed varieties, it had financed the installation of low-cost, labor-intensive wells in the Comilla area by the KTCCA. The ADC also had close connections with the private sector with which it was frequently associated in the provision of agricultural supplies. The ADC took the initiative in competing for authorization and funds for tubewell installation in the late 1960s, when it was informally encouraged by the World Bank to prepare a loan request for submission to the Bank for an initial tubewell development program.

Hopes of meeting the Fourth Plan targets for investment in tubewells depended on foreign aid. Of the Rs. 80 million budgeted for the first year of the plan, Rs. 44 million was to come from foreign aid.[15] The expectation was based on commitments or expressions of interest in tubewells by a variety of donors.

Because the appropriate technology was not yet determined, most decisions were deferred by donor agencies until the World Bank could take the lead. This was in part because the Bank had signed an agreement with East Pakistan in 1967 in which it made the province a special priority area and agreed to provide substantial development policy assistance as well as development loans. This agreement and the fact that the Bank was the organizer of the Aid to Pakistan consortium made it the leader among the aid agencies. Although the government of East Pakistan initiated requests for support of its tubewell program with several aid donors, a response to them was deferred until the Bank could appraise the application submitted to it and establish its policy.

In February 1970, the Bank's tubewell appraisal team, consisting of one British and one American engineer, an economist, an agriculturist, and a representative of the Swedish aid agency, spent three weeks in East Pakistan preparing the report which contained recommendations regarding possible support of the tubewell program by the Bank and the type of technology it preferred. Working against a deadline and thus unable to analyze each alternative in great detail, the team came to an agreement with ADC to finance 3,000 medium-cost wells. Once this precedent had been set other aid donors followed quickly.

The next one was the British government, which provided a guarantee for a supplier's credit from the General Electric Co. (GEC) to provide 5,000 medium-cost wells. The third agreement was reached with Geotechnica, a Yugoslav drilling firm, to install 1,000 medium-cost wells under a Pakistan-Yugoslav barter trade deal. These arrangements were under the auspices of ADC. In late 1970, the World Bank agreed to evaluate a proposal prepared by WAPDA for an extension of the Thakurgaon project with high-cost wells, but WAPDA was unable to obtain financial assistance to support the remaining 1,135 wells it proposed.

By late 1970 the trend of tubewell development was clear. The government requested aid for about 1,500 of each of the three types of wells. In response, commitments were made to finance 9,000 medium-cost and 385 high-cost wells, indicating a strong donor preference for the medium-cost well and a rejection of the low-cost well. (See Table 2.)

In view of the foreign aid donors' preference for medium-cost, capital-intensive wells drilled by foreign contractor over the low-cost,

TABLE 2

East Pakistan Requests and Donor Commitments for Installation of Tubewells in 1970

Type of Well	*Requests*	*Commitments*
Low-cost (under Rs. 50,000)	1,500[1] (ADC only)	none
Medium-cost (Rs. 50,000 to 100,000)	1,500[1]	9,000[2]
High-cost (over Rs. 100,000)	1,510 (WAPDA only) 3-cusec wells	885[3] 285 3-cusec wells 600 1/2-cusec wells

Notes:

1. The application to the World Bank was for 1,000 wells drilled by labor-intensive methods, 2,000 by power, and half of the 3,000 to have low-cost pumps and screens and half medium-cost materials. The net result was that 1,500 would have been low-cost and 1,500 medium-cost.

2. Includes wells financed by the World Bank, General Electric Co., and Yugoslavia.

3. Of the original request for 1,510 wells the World Bank tentatively agreed to finance the equivalent of 385 of these wells in 1971; 285 to be 3-cusec wells and 600 small 1/2-cusec wells (equal to 100 3-cusec ones). Although the 1/2-cusec wells cost much less than Rs. 100,000 per well, the cost per cusec of water means that this figure must be considered high. No aid donors were found for the remaining 1,125 wells requested.

labor-intensive technology in a capital-scarce labor-surplus economy—a preference shared or at least acquiesced in by the government of East Pakistan—it is necessary to analyze the choice more carefully.

III. AN EVALUATION OF TUBEWELL ALTERNATIVES: DEVELOPMENT GOALS

A. *Economic Returns*

The cost of the three types of wells both in 1970 East Pakistan costs and in accounting prices is shown in Table 3. For each of these alternatives internal rates of return* have been calculated. Details of the calculation of benefits and costs are provided in Appendix A.

* In the process of project analysis an internal rate of return is often calculated to determine the discount rate at which costs equal benefits. A comparison of these rates provides a basis for a preferential ordering of projects.

TABLE 3

Comparison of Cost of Three Tubewells of Varying Specifications in Market and Accounting Prices. All figures in Rupees. Accounting Prices shown in parentheses.

		Low Cost		*Medium Cost*		*High Cost*	
	Drilling —	Jet/Percussion		Contractor/Power		Contractor/Power	
	Screen —	Brass		Fiberglass		Fiberglass	
	Pump —	Centrifugal		Turbine		Turbine	
	Engine —	Low speed diesel		High speed diesel		Electric (inc. generation/ transmission cost)	
Moving equipment to and from drilling site		300	(150)	4,500	(4,500)	4,500	(4,500)
Drilling cost							
160 feet				7,600	(13,300)	7,600	(13,300)
180 feet		1,440	(720)				
Pumphousing 10″ diam.							
80 feet				4,800	(9,600)	4,800	(9,600)
Pipe 8″ diameter							
20 feet		800	(1,600)	—		—	
Slotted Screen Pipe							
80 feet				4,280	(13,482)	4,280	(13,482)
140 feet		7,000	(11,900)				
Bail plug and reducer		500	(1,000)	500	(1,000)	500	(1,000)
Gravel pack							
80 feet				1,200	(1,200)	1,200	(1,200)
140 feet		2,400	(2,400)				
Install well hardware							
160 feet				1,920	(1,920)	1,920	(1,920)
180 feet		2,160	(1,080)				

Develop and Test Well	1,500	(750)	1,500	(1,500)	1,500	(1,500)
Pump	750	(1,125)	5,500	(11,000)	5,500	(11,000)
Engine (20 h.p.)	4,500	(6,750)	6,000	(12,000)	125,000	(215,000)
Right angle gear drive	—		1,500	(3,000)	1,500	(3,000)
Pumphouse	3,500	(3,500)	4,500	(4,500)	3,500	(3,500)
Install pump and engine	750	(375)	750	(750)	750	(750)
Consultants	—		5,000	(10,000)	10,000	(20,000)
Field Distribution System	3,500	(1,750)	3,500	(1,750)	5,000	(5,000)
Contingency	2,560	(2,560)	4,955	(4,955)	17,255	(29,975)
TOTAL (in Rs.)	31,660	(35,660)	58,005	(94,457)	194,805	(334,727)

The internal rates of return are as follows:

	Actual Prices	Accounting Prices
Low-cost	48%	54%
Medium-cost	33%	25%
High-cost	7%	4%

On the basis of these calculations, the low- and medium-cost wells are clearly the logical choice, for either public or private investment, although the low-cost ones provide a decidedly more attractive return.[16] The high-cost wells appear to have little economic justification. Since they also lack donor support, subsequent analysis will focus primarily on low- and medium-cost wells.

B. The Effect of the Rate of Installation on Returns

The preceding calculations of returns for different types of wells assume that wells can be installed at the same rate. Most advocates of the contractor-drilled (medium- and high-cost) wells claim that their much higher cost is justified by much more rapid rates of installation. One adviser to WAPDA suggested this in an official report:

> The question is whether or not local contractors can cope with the great number of wells which must be installed in East Pakistan in the next five to ten years. If the capacity is not available and foreign contractors must be brought in, the cost price per unit is likely to increase considerably. However, the increase in cost price may be offset by the total return in the larger number of units which can be installed with the aid of foreign contractors.[17]

To test this assumption, two theoretical alternative Fourth Plan tubewell programs are established for the period 1970-75, a low-cost and a medium-cost tubewell program. Both programs could achieve the Fourth Plan well installation target of 20,000 wells. The low-cost program will do it for Rs. 696 million less or about half the cost, but the medium-cost program will install wells at a faster initial rate. Foreign contractors can install 4,000 wells per year, that is, 20,000 between 1970 and 1975. This has to be compared with the rate at which the capacity to install low-cost wells can be developed. Table 4 shows the maximum expansion rate of the capacity to install low-cost percussion-drilled wells over a five-year period.* (With low-cost techniques 32,784 wells could be

* Low-cost wells are installed by a driller and assistant driller who hire and direct a group of 10 to 20 local unskilled laborers. One foreman supervises the work of five rigs. East Pakistan could have started with 85 of the 103 trained crews

installed for the Pakistan Fourth Plan allocation of Rs. 1,400 million within that period.) Figure 1 compares the two alternative rates of installation. It indicates that more wells can be installed in the first four years by foreign contractors, but by the end of the fifth year capacity to install wells by low-cost methods will have vastly exceeded foreign contractor capacity.

The two alternative programs achieving a similar goal of 20,000 wells in five years but with different rates of installation and different cost are evaluated by rate of return analysis. Compared in this manner, the rates of return were as follows:

Low-cost jet and percussion drilled wells	55%
Medium-cost contractor, power-drilled wells	24%

Interestingly, the lower costs just about perfectly offset the longer time it takes to construct the wells, with the result that the difference between the rates of return of low- and medium-cost technologies is similar, whether comparing two wells installed simultaneously or multiple installation well programs which introduce the time factor. One must therefore conclude that the economic return on the low-cost wells, despite lower initial rate of installation, is substantially higher. This analysis refutes the argument that the time saved by foreign contractors provides the medium-cost wells with a higher economic return. On the contrary, it shows that the low-cost wells provide a better economic return despite an initial lag in installation capacity.

C. *Employment and Training*

In East Pakistan's labor surplus economy, the creation of new employment opportunities ranks high. To sink a well by percussion or water jet methods, two shifts of eight unskilled laborers are employed for a four-week period, a total of 16 man-months or 1-1/3 man years. If

TABLE 4

Rate of Increase of Well Installation Capacity
by Low-Cost Methods

Year	*1971*	*1972*	*1973*	*1974*	*1975*
Rig operation	85	212	530	1,325	3,312
Annual capacity	510	1,272	3,180	7,950	19,872
Cumulative installation	510	1,782	4,962	12,912	32,784

available in 1970. Given the capacity of each crew to train three new assistant drillers annually, who can become drillers or foremen after additional work experience, one rig could train the personnel needed for 1½ new rigs each year. Additional detail is provided in Appendix B.

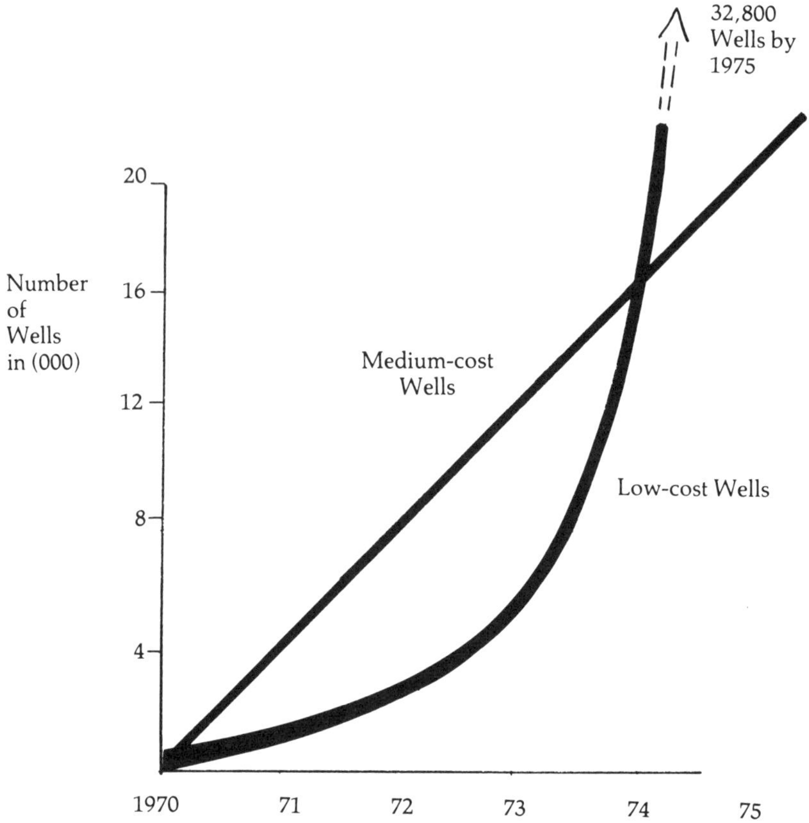

Figure 1—Comparison of Rate of Installation of Wells by Low and Medium-cost Drilling Methods

20,000 wells (the Fourth Plan target) of this type are installed, more than 26,000 man-years of employment for unskilled labor would be created. In addition, employment would be created for the drillers and assistant drillers. This would amount to another 9,000 man-years of employment, or a total of 35,000 man-years for the program. By contrast, the medium-cost, power-drilled wells employ far fewer people. The 160 power rigs needed to drill 4,000 wells annually would employ two men each, amounting to 1,600 man-years employment for 20,000 wells. Therefore, in terms of the employment objective of East Pakistan, too, the low-cost wells were decidedly preferable.

In a society in which the absence of skilled manpower shackles development, a program of tubewell development that would train 7,286 foremen and drillers is clearly preferable to employing foreign contractors whose operations would provide training to only a few hundred Bengalis. In addition, percussion and jet drilling would provide 524,500 laborers with two to four weeks experience in drilling operations. In terms of training, the low-cost system is clearly preferable.

This is not to dispute the fact that foreign contractors can make an important contribution to the general knowledge of tubewell technology, particularly in the early stages of the program. They possess much greater experience and greater engineering knowledge and they can impart important training in the techniques and skills of well installation. Such training is a benefit in depth because it increases the capacity of local well technicians instead of providing broad training in simple mechanics and well technology to a wide range of rural people as do the low-cost wells.

D. Location and Distribution of Benefits

Both the World Bank and the Yugoslavs (presumably GEC, too, although details of the agreement are not known) decided to limit their drilling sites to the same six of the nineteen districts of East Pakistan, mainly because these are the areas in which the heavy ten- to fifteen-ton power rigs used by foreign contractors for the medium-cost wells could be moved easily from one drilling site to another. In other areas, mobility of heavy power rigs is severely limited by the numerous rivers and canals and the absence of roads and bridges. The low-cost jet and percussion drilling rigs, however, weigh relatively little, and can be easily taken apart and reassembled. As a result, they can be transported from one site to another by almost any means, tractor-drawn wagon, boat, animal-drawn cart, or even men, and can therefore be operated almost anywhere in East Pakistan in any season. This mobility makes the low-cost drilling equipment much more versatile than medium-cost equipment and thus its benefits can be more broadly distributed across the country.

Moreover, the installation of low-cost wells provides a broader distribution of employment, training, and financial benefits, most of which are retained in the rural areas. On the other hand, the benefits from medium- and high-cost power-drilled wells go to city-based or foreign contractors, and the installation—in itself—provides almost no benefit to the location in which it takes place.

E. Linkage Effects

Development of small industry (some of it rural) would be an important by-product of a tubewell program. With some encouragement from the government, the number of private domestic contractors installing wells would inevitably increase, no matter what technology is chosen. The important question, then, is what local small industries would be developed by alternative techniques. Since the medium-cost program was based almost entirely on imported materials while the low-cost program would utilize supplies produced primarily in East Pakistan, the low-cost program was much preferable from the perspective of promoting local small industry. The magnitude of such benefits from a low-cost tubewell program is discussed below.

Rigs. The components of percussion and jet drilling rigs could all be fabricated in East Pakistan. By the fifth year, Rs. 1.5 million would have been spent domestically to obtain the anticipated 3,000 rigs. Their fabrication and maintenance would have provided the basis for one or more small firms.

Pipe and Screen. If the average well is 160 feet deep as estimated, 20 feet of pipe and 140 feet of screen would be required. In a 20,000 well program, there would be demand for 2.8 million feet of screen and 400,000 feet of blind pipe. This would mean an expenditure of Rs. 140 million ($30 million) on screen and Rs. 20 million ($4.5 million) on blind pipe. For low-cost wells, blind pipe was manufactured in West Pakistan and brass screen in East Pakistan in a relatively simple process which could be expanded to additional small workshops around the country.

Pumps and Engines. Centrifugal and turbine pumps and low-speed diesel engines have been manufactured in Pakistan. Supplying these for a low-cost program would have meant Rs. 105 million of investment in domestic industry.

As these figures make clear, the low-cost program would have provided a major stimulus to small industry in East Pakistan. In terms of East Pakistan's development goals, economic growth and its distribution and employment creation, the low-cost technology was greatly preferable to the medium or high-cost techniques. If the decision were based on these considerations alone, there could be no question as to the preferable technology.

IV. AN EVALUATION OF ALTERNATIVE TECHNOLOGIES: ORGANIZATION AND IMPLEMENTATION

In addition to the development goals of East Pakistan, questions of organization and implementation were of paramount importance. In an area like East Pakistan with a very limited reservoir of trained administrators attention had to be given to finding an efficient organization for sinking wells. Since in numerous irrigation projects, including Thakurgaon, tubewells yielded limited benefits partly because of failure to train and organize farmers to utilize water, this factor too had to be considered in the choice of technology.[18]

A. Installation Capacity

Capacity to install wells according to a predetermined schedule of costs and specifications was a high priority. Low-cost wells as proposed by ADC were to be installed under the corporation's auspices. This technology was also suitable for use by local organizations. Medium-cost wells under the World Bank's proposal were to be installed by contractors. Contractors were to be selected by bidding. By the Bank's decision, bidding on 2,100 wells was to be open only to foreign contractors, but either domestic or foreign contractors could bid on the remaining 900.

The Bank believed that, given the limits of local administrative capacity, foreign contractors were apt to prove more reliable in installing wells on schedule and to agreed-upon specifications than the ADC. This appears to be true even though the past performance of foreign contractors in East Pakistan, admittedly under highly variable natural conditions and in a somewhat unpredictable bureaucratic atmosphere, suggests that foreign contractors have had serious difficulties meeting their formal commitments.[19] Yet under these circumstances, in which both the ADC and a foreign contractor might encounter difficulties, the foreign contractor appears to be the better risk, and this was an important factor in influencing the Bank's choice.

B. Organization of Demand for Water

The failure of past irrigation projects in East Pakistan resulted in part from an absence of farmer demand for water. Until 1967, irrigation projects focused on the supply side and neglected demand for water. With the Low lift Pump Program utilizing surface water, which made demand a prerequisite for supply, an important start toward solving this problem was made.

Demand and the capacity to organize irrigation groups are the critical determinants of the rate at which the number of tubewells can be expanded. If the rate of installation surpasses demand, then the rate must be slowed because wells sunk in excess of demand will be only partially used. Although it is difficult to determine in advance the rate at which demand for water will expand, it is clear that once available and utilized, irrigation water becomes a highly prized commodity which villages will fight to retain.[20]

At Comilla, the demonstration effect was important in determining demand, and demand regulated the rate of expansion. Wells were installed only after irrigation groups were formed, applications submitted, and down-payments made. In the first three years of the program, only 35 wells were installed. By the fourth year, the rate of demand and of installation approximately doubled, and in the fifth year, 54 wells were installed and there were applications for over 400 more. These figures suggest that the demonstration effect may cause a lag in demand and that the installation of wells at a high rate in the early years may not be desirable. This lag in demand may well coincide with the time needed to develop capacity for installation of wells by low-cost techniques.

It is quite probable that installation techniques will have some influence on farmer utilization, operation, and maintenance, as they did in Comilla. Although there is no way to quantify this psychological aspect, farmers who have participated in the installation of a well, having spent four weeks working on it, are likely to have a much greater understanding of well operations as well as a feeling of personal investment in it. As a result, they may be much more aware of its potential benefit. If, on the other hand, the process is one of bringing in heavy equipment which they do not understand and which in one week installs the well and departs, they will not understand how the well works or feel it is anything with which they are associated. Thus the farmer's utilization prospects appear better with low-cost wells.

C. Potential for Non-governmental Installation

Although potential for private-sector investment in tubewells was limited for reasons explained earlier, some existed in East Pakistan. Because of the area's needs, it was important that this demand be met despite the fact that the government, as we have seen, put much higher priority on public-sector tubewell development. Clearly, the low-cost technology with its higher returns would have been preferred by the private investor. To encourage private investment, the government might have had to stimulate contractors with this capacity by offering a few government contracts for installation of low-cost wells.

In the 1960s the Comilla program was based on local cooperatives as implementing agencies for tubewells. Since the effective system of farmers' cooperatives developed at Comilla was scheduled for expansion in East Pakistan, it would have seemed logical to try to take advantage of the model established by this one successful tubewell project to tie tubewell expansion to the expanding cooperative system. Only the low-cost technology seemed appropriate for this purpose.

On balance the arguments for the low-cost wells over medium- and particularly high-cost were impressive. East Pakistan's development objectives, economic return, employment creation and training, the distribution of benefits as well as the potential for the creation of domestic industry would have been better served by the low-cost wells. The relative merits in terms of organization and implementation are less clear-cut. The belief that foreign contractors were more reliable was an important factor in favor of medium-cost wells, while the advantage of low-cost wells in terms of farmer utilization of water and the private sector preference for low-cost wells was in their favor. On balance, this evidence, plus the fact that the low-cost wells were the only ones proven in actual operation in East Pakistan suggest that the low-cost wells utilizing the simple technology represented the tubewell technology most consistent with the objectives and needs of the country. The fact that the government and the World Bank rejected this technology suggests that standards other than those examined played a prominent role in the decision.

V. FACTORS AFFECTING THE CHOICE OF TECHNOLOGY

The choice of the medium-cost technology is puzzling on first examination. The World Bank, for reasons already indicated, as well as the government were strongly committed to the development objectives of East Pakistan. Considerations of administrative feasibility did not offset these concerns. Yet a technology that was definitely not optimal was chosen. Since the economists, engineers, and administrators who participated in the decision presumably shared East Pakistan's development objectives, it is necessary to look at other factors affecting the choice.

Perhaps the central characteristic of the procedure by which such a decision is arrived at is that, instead of being a process intrinsic to an institution or governmental system, it is a bargaining process between the government and an external institution. When this process crosses national and cultural lines as well, the difficulties of arriving at an agreement on an optimal solution are greatly increased. The availability of external aid to finance new investment in itself affects the choice of technology, for the preference of the aid giver then becomes an important element in the decision-making process. The administrator in a developing nation may very rationally accept a technology that he considers second or third best if foreign financing is available only for that choice, since that may be the only way to receive this aid. Furthermore, the official may tailor his program to the technology he considers most likely to attract foreign aid.

On the donor side, many considerations other than the needs of the recipient country may affect the type of aid or credits made available. In fact, the form of aid frequently depends more on the requirements of the donor country's economy than on those of the recipient's. Clearly, national commercial interests affected the forms of assistance available under the GEC supplier's credit arrangement and the Yugoslav barter agreement. An international agency like the World Bank, however, is not directly tied to parochial interests or to profit-making objectives; nor are the developing countries seriously constrained from arguing their own interests. Yet both the Bank and East Pakistan concurred in a tubewell technology that was less than optimal for East Pakistan. The following discussion of economic factors, various perceptions of the issue, and organizational objectives provides some additional insights into the influences upon the tubewell decision.

A. Economic Factors

Economic factors can play an important role in the choice of technology. Price distortions, caused by inconsistencies in the tax and duty structure, may influence choices as they did in East Pakistan where the market cost of imported high-speed diesels was less than that of the locally produced low-speed engines, even though the latter cost only 75% as much to produce. Similarly, the contractors of government agencies implementing tubewell programs will have a rational preference for scarce capital goods (machinery) over abundant human labor when selecting technology, if the local currency is overvalued, as was the Pakistan rupee. In that situation imported equipment, such as pumps or power drilling rigs, could be obtained for as little as half their true cost to the economy. The result is that the price of capital goods is subsidized, a benefit which any economizer, public or private, is happy to accept.

If institutionally established wage rates exceed the real cost of labor, as they did in East Pakistan in 1970, labor-intensive methods become less attractive. When the problems of management of labor crews are added, contractors will generally adopt capital-intensive methods despite the fact that this is inconsistent with the nation's need to utilize low-cost labor.

Although economic factors can bias the choice of technology, project analysis techniques such as accounting prices can offset these factors. The rate of return analysis on p. 44 demonstrated that despite subsidized capital, the low-cost technology yielded a better economic return.

B. Alternative Perceptions of the Issues

Participants in a decision may have different perceptions of the same problem. Foreigners new to the country may perceive the choices very differently from economists, and they both may see things differently from the administrator or the farmer.

For the World Bank, emphasis in this initial tubewell program in East Pakistan was on quality and reliability, concerns suggested by the past experience of the members of the appraisal mission and their institutional outlook. Their experience led most in the mission to favor reliable, imported technology. One of the Bank's engineers who was appraising tubewells in Comilla remarked while watching the low-cost wells being installed in a sea of mud by a large group of villagers, "You can't install reliable tubewells this way." Yet tests and specifications showed that the labor-drilled wells were of a quality equal to the power-drilled wells.

The reliability of the wells had to be judged both in terms of the tubewell itself and of the environment in which the wells had to operate. To most of the Bank's appraisal staff, the power-drilled, medium-cost wells appeared to be more reliable. Yet, although it is evident that the preferred, high-speed, diesel engine may operate longer and more reliably in ideal conditions, in an East Pakistan village, where there are few trained operators or mechanics, it is less tolerant of misuse and much more difficult to repair than the low-speed engine. The well that may be inferior in technical and engineering terms may prove more reliable for the farmer, more suitable for the conditions in which he operates, and therefore more efficient in producing the ultimate objective of increasing agricultural output.

An evaluation of a well's efficiency has another dimension. Many technicians fail to see the step beyond efficient water production: the construction and economical maintenance of wells throughout the countryside. The critical issue concerns not only the type of well that produces water efficiently but also the type that has a mixture of cost and performance characteristics as well as maintenance potential that makes it suitable for widespread use in the area. Examples from both West Pakistan and Vietnam illustrate these perceptual problems. Prototypes for low-cost tubewells in West Pakistan and for low-lift pumps in Vietnam were dismissed by engineers as having "little technical validity" after they had been tested and proved inefficient (in an engineering sense). But in each case, the innovative prototypes proved to be of a cost and specification that made them extraordinarily profitable under the existing conditions and their use spread very rapidly.[21]

Risk is another consideration that was perceived very differently. For the farmer, the risky technology was that which was installed by outsiders and which he could not operate or repair, and therefore might not use. The simple technology was more adaptable to changing conditions. In case of failure, much less was invested in an individual unit. The one successful pilot project—in Comilla—had been carried out with this technology. For the World Bank, the reliable installation agent was the foreign contractor, despite the fact that the pilot tubewell project at Thakurgaon carried out by foreign contractors was largely a failure. For the Bank, concentrated drilling locations where they could observe operations implied less risk than a decentralized, locally supervised operation.

Finally, there was the view held by some Pakistani officials in which anything "modern" was preferable. The appearance of modernity was important to this group; to them it was synonymous with "the best" and the best was what in their view aid donors should provide.

These varying perceptions all contained important elements of reality. They led people to differ honestly as to the appropriate tubewell technology for East Pakistan. In this case, the role of the aid donors was particularly influential and undoubtedly contributed to the choice of medium-cost technology.

C. *Organizational Objectives*

The search for an explanation of why organizations behave in ways that appear contrary to their stated objectives would strike a student of organization theory or decision-making as very familiar. Indeed, much of the literature of administration, beginning with Chester Barnard, has touched on this issue, for people in organizations frequently act in ways that appear irrational.[22]

In this case, the organizational requirements of the developing countries and the aid donor agencies appear to have superseded development policy objectives. Thus, organization theory provides more insights into the choice of technology than does economic theory and proves useful in explaining the choice of tubewell technology.

Work in this field has stressed the concept of "bounded" or limited rationality in decision-making. According to the theory, the limits of man's capacity as solver of complex problems circumscribe the decision-making processes of individuals and organizations by forcing them to simplify problems in a variety of ways which result in replacement of the goal of finding the optimum solution with that of finding a course of action that appears satisfactory to them.[23] Elaboration of the factors that limit the exercise of full rationality helps to explain the choice of tubewell technology.

1. Satisfactory, Rather than Optimal Solutions

Since an optimal solution is rarely identifiable or ascertainable, a satisfactory one is accepted. In this case, established procedures, protection of bureaucratic domains and routines, and the need to avoid risk were all obstacles blocking the economically rational solution. In addition, limitations on information and time prohibit the decision-maker from considering all possible solutions to a problem and force him to choose among the first satisfactory alternatives. The World Bank

Appraisal Mission was very short of both information and time. It was not able to make all the calculations or obtain all the information contained in this paper. As a result, the members chose the solution that seemed to fit their perception of what was needed, met their organizational requirements, and also had a satisfactory economic justification. The medium-cost technology was familiar and was considered reliable. Since the World Bank was able to show an anticipated 32% internal rate of return for their medium-cost wells, a less than optimum solution was satisfactory.[24]

As far as the government officials were concerned, aid was available only for medium-cost wells, and this technology conformed with their institutional requirements and perhaps their personal preferences as well. For both the Bank and the government, a tubewell project on which they could cooperate was a satisfactory basis for action.

2. Avoidance of Uncertainty and Reliance on Established Programs of Activity

Minimization of risk characterizes organizations and individuals within them. This tendency is a deterrent to innovation and stresses repetition of procedures that are familiar and tested. For the Western engineer, machine power drilling, fiberglass screens, high-speed engines, and turbine pumps represent a familiar and reliable technology unlike the low-cost technology. Foreign drillers who could be held contractually responsible for performance appeared safer (despite serious mistakes at Thakurgaon) than a decentralized program involving a large number of low-cost rigs. Thus, familiarity and the avoidance of risk were preferable to optimum results.

For government officials, technology with the appearance of modernity is less subject to criticism than simple technologies. Furthermore, they could reduce personal responsibility by claiming they had accepted less than optimal technology under pressure of aid officials in order to obtain aid.

3. Operating Procedures and Established Routines

Organizations are bound by their operating procedures and established routines. Because an agency like WAPDA is staffed with engineers trained in modern construction methods, supported by large groups of foreign consultants and a flow of foreign aid, has a large, well-capitalized construction division, and, finally, makes decisions on the basis of feasibility studies, it is almost incapable of employing low-cost technology. Its staff, equipment, and procedures are all oriented towards high-cost, high-quality construction. Although the other East Pakistan sponsor of tubewell programs, the Agricultural Development Corporation, was less tied to "high" technology, the head of their water development division had just come from WAPDA and he and others had strong preferences for a "modern" technology. For both the Bank and the government, existing organizational routines and physical procedures limit the options available to them in the short run.

4. Control

Control is an objective of most organizations. It is closely related to risk minimization and was an important factor both for government agencies and the World Bank. Government agencies and their staffs derive power, prestige, and sometimes an opportunity for profit by attracting foreign aid and implementing large programs. For the Bank, control was equally important to ensure proper implementation to protect against misuse of funds, and to ensure maximum return on the project. These objectives were, in the Bank's eyes, best accomplished in a program it could supervise closely and control. For government and Bank alike, a low-cost program with up to 3,000 rigs operating at scattered locations throughout the country required a decentralized administrative system and a resultant loss of control. To avoid this, they were all willing to accept less than optimal economic returns and social benefits.

Unfortunately, most of these organizational factors were implicit, not explicit. Most of the participants in the decision-making process contended they were acting only on the basis of development goals and administrative feasibility, although the evidence seems clear that they were not. The result was that there was no conscious effort to weigh all the real considerations and the trade-offs among them. Had such a weighing of trade-offs been possible, solutions to organizational goals might have been found that would have permitted the most desirable technology to be chosen. The low-cost wells might have been restricted to a tubewell field where closer supervision would be possible, thus satisfying the Bank's need for supervision and control. With more time a different foreign contractor might have been found who could deal with the low-cost technology, or at least part of the funds might have been allocated as risk capital for experimental wells that could test the merits of alternative technologies, and perhaps even develop new alternatives that might be preferable to those considered here. Much would have been gained if legitimate organizational objectives had been made explicit. Then consideration of the trade-offs between these and the more explicit development goals might have led to the selection of an optimal technology.

Ultimately, it was the organizational requirements of the implementing agencies, including the aid donors, that determined the choice of tubewell technology for East Pakistan. In the actual decision-making, such factors as risk avoidance, appearance of modernity, established procedures, familiar techniques, and by no means least, control, outweighed development policy objectives. It is in these factors that an understanding of decisions as to choice of technology must be sought.

Appendix A

NOTE ON CALCULATION OF INTERNAL RATES OF RETURN ON TUBEWELLS

A variety of techniques can be used for appraising the economic returns of projects. In this case, internal rate of return was employed because it was used by the World Bank in its analysis of tubewell alternatives for East Pakistan. This makes possible direct comparisons with the Bank's calculations, and provides an easy way to compare two alternative programs with different life expectancies. While other techniques could have been used, they would not have served these two purposes as well.

1. Costs. The details of market costs of tubewell components are provided in detail in Table 3 and need no further explanation here. The accounting pricing of foreign exchange and labor do need comment. The official rate in 1970 was Rs. 4.75 = $1. The accounting exchange rate was established at twice that, Rs. 9.50 = $1. This was done, again in part because it was the rate used by the World Bank in its well evaluation. Its choice of an accounting rate was similar to the free market rate of the rupee during that period.

The accounting wage rate was also an approximation. The regular market rate for unskilled labor was Rs. 2.50 to Rs. 3.50 per day. Since widespread underemployment was an established fact that is partially documented in the paper, an accounting rate of Rs. 1.50 per day was used. Because labor costs were not a major factor in installed well costs—even those installed by labor-intensive methods—the results are not substantially altered by an accounting rate. In fact, there is only a 1.5% increase in the rate of return on labor-intensive wells when labor is accounting priced.

2. Benefits. It was not possible to include all the benefits from irrigation in these calculations. First, it proved impossible to quantify with sufficient accuracy benefits from supplemental irrigation of *aus* and *amon* rice crops, so although they may be substantial, they are not included. Second, rice is one of the most expensive crops to grow on irrigated land because of its high rate of water consumption. It can be assumed that as irrigation becomes more widespread and rice production increases toward levels of self-sufficiency, cropping patterns will change

and higher value crops will be grown in the irrigated season, resulting in higher returns on each well. Third, because of difficulties in finding a high-yielding rice that is entirely appropriate to the conditions of the winter season, the irrigated crops planted between 1968 and 1970 were partially new varieties and partially traditional varieties. Therefore these benefit calculations are based on the assumption that half of the irrigated area will be planted in high-yielding varieties and half in traditional varieties.

In calculating benefits, the experience in Comilla was utilized because this is the only area from which detailed information regarding the cost and benefits of irrigated agriculture was available. For the net benefit (yield minus all inputs except irrigation) per acre of rice at market prices, the average wholesale price of the first three months of 1970, Rs. 30.80 per *maund* (82.7 lbs.), was taken as the standard value of rice. Table A-1 contains the calculation of benefits from one acre of irrigated land planted in high-yielding varieties and one acre planted in traditional varieties. Assuming that one 2-cusec tubewell will irrigate 35 acres, which was the average for a 2-cusec water source in East Pakistan in 1970, it is then possible to calculate the net annual benefit accruing from a 2-cusec tubewell, assuming the cropping pattern of half high-yielding and half traditional varieties of rice. Table A-2 summarizes these calculations.

This calculation is important because it suggests the profitability of a tubewell to the water users, a profitability sufficiently high to repay the investment cost of a low-cost well in less than two years or even a medium-cost well in three years!

Since costs were calculated at accounting prices, it is necessary to use the same procedure for benefits where increased production replaces imports. The real value of additional rice in a food grain importing area like East Pakistan is equal to the amount of foreign exchange saved by the domestic production of each additional ton of rice. For this purpose the 1970 export price for Burmese rice, $103 per ton is used[1] and this is converted to rupees at the accounting rate of Rs. 9.50 = $1. (The profitability of wells is underlined by the fact that by mid-1973 the international price had risen to $196 per ton.[2]) Calculated at the international price, the accounting benefit from one tubewell increases to Rs. 32,633 per well, as against Rs. 22,197. If a turbine pump is used, with a resulting 25% increase in output, the benefit rises to Rs. 40,791. Calculations on medium and high-cost wells utilize the two higher benefits.

3. Sensitivity Analysis. The calculations were tested for sensitivity to changes in a number of the variables. In the cost calculations, shifts in either the shadow price for foreign exchange or for labor, within

TABLE A - 1

Calculation of Net Benefit per Acre of Tubewell Irrigated Crops*

(costs in Rupees)

Improved Varieties of Rice

Costs per acre

Input cost, except irrigation (labor, power, fertilizer, seeds, insecticides)	Rs. 343.00
Overhead costs (interest on land and credit, land and sales tax)	Rs. 110.43
Total cost per acre	Rs. 453.43

Benefits per acre

	Yield (mds.)	*Price per md.*	*Return per acre*
Grain	41.3	30.80	1,272.04
Straw	19.4	3.00	58.20
			1,330.24

Net benefits per acre: Rs. 876.81

Local Varieties of Rice

Costs per acre

Input costs	Rs. 236.53
Overhead costs	Rs. 84.38
Total cost per acre	Rs. 320.91

Benefits per acre

	Yield (mds.)	*Price per md.*	*Return per acre*
Grain	21.8	30.80	671.44
Straw	13.7	3.00	41.10
			712.54

Net benefits per acre: Rs. 391.63

*Source: Anwarul Hoque, *Cost and Returns*, (Comilla, Pakistan Academy for Rural Development, 1968).

reasonable ranges, made no more than two percentage points difference in the rate of return.

The calculation of rates of return did prove sensitive to changes in benefits. Given the conservative nature of the benefit estimates, almost any change would increase the rate of return. Such change would include the number of acres irrigated per well, the proportions of high-yielding to traditional rice varieties, or the price of rice. For example, a 10-acre increase in area irrigated per well would increase the accounting rate of

TABLE A - 2

Calculation of Benefits Accruing From One Tubewell

Crop	Net Benefit Per acre (rupees)	No. Acres Per 35 Acre Pump Group	Net Annual Benefit Per Tubewell (rupees)
Improved variety rice	876.81	17.5	15,344.18
Traditional variety rice	391.63	17.5	6,853.53

Net annual farmer's benefit Rs. 22,197.71
(excluding irrigation costs)

return on low-cost wells to almost 75%. If the price of rice were to go to Rs. 50 per *maund*, the rate of return would have been over 100%. These tests of sensitivity emphasize that tubewells were extremely profitable, since there seem to be few factors that would lower the benefit and make returns lower than those used in the paper.

One final test was made, however, and that was to see how delays in implementation would affect the comparisons of low- and medium-cost wells at accounting prices. The test used is a rather imperfect one, but the best available. Instead of deferring benefits one year behind costs as was the standard in all these calculations, the benefits for low-cost wells were deferred two years. This lowered the rate of return to 36%, considerably less than the 54% return without the extra year lag, but still preferable to the 25% return on medium-cost wells at shadow prices.

4. Comparisons of wells of alternative specifications. Table A-3 shows the time streams of benefits and costs for the comparison of the internal rates of return on low-, medium-, and high-cost tubewells. The sources of costs for these well alternatives at market and accounting prices are contained in Table 3 in the text.

5. Rate of return for alternative time sequences of installation. The calculation shown in Table A-3 compares the three well alternatives open to East Pakistan in a static situation. The following calculations attempt to compare two alternative well installation programs that might have been undertaken between 1970 and 1975. For purposes of the calculations, it is assumed that the low-cost wells have brass screens with a ten-year life and a combination of centrifugal and turbine pumps. The medium-cost wells are assumed to have a fiberglass screen with a twenty-year life, and turbine pumps. The installation of power-drilled wells is to take place at the rate of 4,000 per year and the

TABLE A - 3

Comparison of Rate of Return on Three Alternative Tubewells of Varying Specifications
(in Rupees)

Year	*Low-Cost* Benefits	*Low-Cost* Costs	*Medium-Cost* Benefits	*Medium-Cost* Costs	*High-Cost* Benefits	*High-Cost* Costs
1	0	31,660	0	58,005	0	194,805
2	22,200	6,500[1]	27,750	8,000[2]	27,750	8,000[2]
3	"	"	"	"	"	"
4	"	"	"	"	"	"
5	"	"	"	"	"	"
6	"	"	"	"	"	"
7	"	"	"	"	"	"
8	"	"	"	21,000[3]	"	21,000[3]
9	"	"	"	8,000	"	8,000
10	"	"	"	"	"	"
11	"	"	"	"	"	"
12			"	"	"	"
13			"	"	"	"
14			"	"	"	"
15			"	21,000[3]	"	21,000[3]
16			"	8,000	"	8,000
17			"	"	"	"
18			"	"	"	"
19			"	"	"	"
20			"	"	"	"
21			"	"	"	"
22			"	"	"	"
Internal Rate of Return		48%		33%	less than	7%
Accounting Benefits/Accounting Costs	32,633 /	35,660	40,791 /	94,457	40,791 /	334,727
Accounting Rate of Return		54%		25%	less than	4%

1 Annual operation and maintenance cost, low-speed diesel engine (estimated life 15 years). Doubled for shadow cost
2 Annual operation and maintenance cost, high-speed diesel engine (estimated life 7 years). Doubled for shadow cost
3 Replacement costs of pump and engine plus operating cost. Doubled for shadow cost.
(Operating and maintenance costs from A.D.C., *Tubewell Loan Application*, p. 113)

percussion-drilled wells according to the schedule in Table 4 (in the text). The results of these calculations are shown in Table A-4 and demonstrate that a program of low-cost wells, despite an initially slower rate of installation would still have a substantially higher rate of return than a program of wells utilizing the medium-cost technology.

TABLE A - 4

Comparison of Rate of Return on 20,000 Wells, Low-Cost and Medium-Cost Wells
(In Accounting Prices, Million Rupees)

Low Cost Wells, Rs. 35,660

Year	Operating Wells	Benefits	Operation & Maintenance	Capital & Replacement Cost	Total Cost
1	510	0.0	0.0	18.2	18.2
2	1,780	16.6	6.3	45.3	51.6
3	4,960	58.1	23.1	113.4	136.5
4	12,910	161.9	64.5	283.5	348.0
5	20,000	421.3	167.8	252.8	420.6
6	"	652.7	260.00	0.0	260.0
7	"	"	"	"	"
8	"	"	"		"
9	"	"	"		"
10	"	"	"		"
11	"	"	"		"
12	19,490	636.0	253.4		253.4
13	18,220	594.6	236.9		236.9
14	15,040	490.8	195.5		195.5
15	7,090	231.4	92.2		92.2

Internal rate of return (using accounting prices) 55%

(continued on next page)

TABLE A - 4 (continued)

Medium-Cost Wells, Rs. 94,457 (In Accounting Prices, Million Rupees)

Year	Operating Wells	Shadow Benefits	Operation & Maintenance	Capital & Replacement Cost	Total Cost
1	4,000	0	0	377.8	377.8
2	8,000	163.2	64	377.8	441.8
3	12,000	326.3	128	377.8	505.8
4	16,000	489.9	192	377.8	569.8
5	20,000	652.6	256	377.8	633.8
6	"	815.8	320	0	320
7	"	"	"	0	320
8	"	"	"	0	320
9	"	"	"	104	424
10	"	"	"	104	424
11	"	"	"	104	424
12	"	"	"	104	424
13	"	"	"	104	424
14	"	"	"	0	320
15	"	"	"	0	320
16	"	"	"	104	424
17	"	"	"	104	"
18	"	"	"	104	"
19	"	"	"	104	"
20	"	"	"	104	"
21	"	"	"	0	320
22	16,000	652.6	256	0	256
23	12,000	489.9	192	0	192
24	8,000	326.3	128		128
25	4,000	163.2	64		64

Internal rate of return (using accounting prices) 24%

Appendix B

CALCULATION OF LOW-COST WELL INSTALLATION CAPACITY

The Comilla KTCCA low-cost wells are installed by a driller and assistant driller who hire and direct a group of 10 to 20 local, unskilled laborers. One foreman supervises the work of five rigs. One rig using either low-cost method can drill about six wells per year.

Drilling rigs are simple and can be fabricated more rapidly than men trained to operate them, so the limitation is the time needed to train crews. In Comilla, six months of apprenticeship on a drilling rig was the time needed to train assistant drillers. After an additional half year's work, an assistant driller is promoted to driller and is given charge of a rig. Foremen are selected from among the drillers on the basis of aptitude, knowledge, and organizational ability.

As of the end of 1970 there were 87 drilling crews which had been trained at Comilla. There were enough foremen and drillers to begin operating 85 rigs immediately. In addition to this, there were at least 14 local contractors with capacity to drill percussion wells. However, private contractor capacity is not included in these calculations.

Since drilling can be done with efficiency in only nine months of the year, it is possible to train only three men on each rig each year. (In other words, one 2-man team in six months and half the training of a 2-man team in the remaining three months.) If you begin from the existing 85 rigs and calculate that each one trains three men per year, it is possible to have 3,312 rigs in operation in five years and to install 32,784 wells in that period.

Notes

1. Government of East Pakistan, *Economic Survey of East Pakistan—1969-1970*, (Dacca, 1970) Table III, p. 104.
2. Government of East Pakistan, *Statistical Digest of East Pakistan*, (Dacca, 1968) Table 3.3, p. 40.
3. Government of East Pakistan, *Economic Survey of East Pakistan—1969-1970*, (Dacca, 1970) Table I, pp. 102-3.
4. *Ibid.*, Table 19, p. 24.
5. Planning Department, Government of East Pakistan, *Strategies and Programme for the Fourth Five Year Plan for East Pakistan*, (Dacca, 1969) p. 124.
6. Based on unpublished estimates made by the Ganges Brahmaputra Basin Study Group of the Harvard Center for Population Studies.
7. During the period 1965 to 1970 *boro* (December-May) rice averaged 18.2 *maunds* (1 *maund* = 82.7 lbs.) per acre. *Aus* (May-July) harvests yielded 10.1 *maunds* and *amon* (August-November) production averaged 12.4 *maunds* per acre. Department of Agriculture, *Agricultural Production Statistics* (Dacca, mimeographed, issued annually).
8. See John Woodward Thomas, "Development Institutions, Projects, and Aid in the Water Development Program of East Pakistan," (March 1972) p. 5 for details.
9. Government of Pakistan, *The Fourth Five Year Plan—1970-1975*, (Islamabad, 1970) pp. 275, 311. After independence, the government of Bangladesh also placed high priority on tubewell development. Its first annual plan for 1972-73 contains provisions for the installation of 6,400 irrigation tubewells of varying depths. Planning Commision, *The Annual Plan, 1972-73*, (Dacca, 1972) pp. 27-8.
10. East Pakistan Water and Power Development Authority, *Irrigation Extension Activities in Thakurgaon Tubewell Project*, (mimeographed, 1969) Appendix A, pp. i-v.
11. Farmer's benefits from a 2-cusec well are estimated at Rs. 22,250 annually, enough to pay the full costs of well installation in two years. See Table A-2.
12. Amit Bhaduri's "Agricultural Backwardness under Semi-Feudalism: A Case Study of West Bengal Villages," (mimeographed, 1972) is an excellent analysis of the interconnection between technology and ownership in land. It provides persuasive evidence that moneylenders in Bengal will not lend to small cultivators for technological improvements, primarily tubewells, because the productivity of these wells will free the small cultivator from dependence on the landowner/moneylender.
13. See John Woodward Thomas, "The Development of Tubewell Irrigation in Bangladesh: An Analysis of Alternatives," (February 1972) pp. 22-43 for more details.
14. John Woodward Thomas, "Development Institutions, Projects, and Aid in the Water Development Program of East Pakistan," (mimeographed, 1972) p. 12.

15. Government of East Pakistan, *Annual Plan 1970-1971,* (Dacca, June 1970).

16. The tubewells financed by the World Bank cost Rs. 13,000 more than the medium-cost wells on which these calculations were made, but by assuming more efficient use of water than in the past and a somewhat higher price of rice, it calculated the rate of return on medium-cost wells at 32%.

17. Peter Mulder, ILLACO, General Consultants to WAPDA, "Socio-Economic Cost of Irrigation by Tubewells and Low Lift Pumps," (Dacca, 1970, mimeographed) p. 6.

18. See John Woodward Thomas, "Development Institutions, Projects, and Aid in the Water Development Program of East Pakistan," for a discussion of problems of farmer organization in six large projects in East Pakistan.

19. *Ibid.* The repeated review and frequent dismissal of foreign contractors working in East Pakistan, some of whom were working on projects funded by the World Bank, underlines the difficulties of working in an area like East Pakistan.

20. The East Pakistan newspapers occasionally carried reports of clashes over water. One report from Mymensingh read, "On the morning of May 9 two groups of people clashed with each other over water pumped out from an irrigation pump in an IRRI paddy field in village Hoglakandi under Hossainpur Police Station. As a result two persons of the locality, Abdul Hai and Abdul Mojeed, lost their lives, it is learnt." *Dacca Morning News,* May 26, 1970.

21. See John Woodward Thomas, "The Development of Tubewell Irrigation in Bangladesh: An Analysis of Alternatives," Appendix A and, particularly, quotations it contains from Aloys Michel, *The Indus Rivers,* (New Haven, Yale University Press, 1967) for documentation of the West Pakistan case and Robert Sansom, *The Economics of Counterinsurgency,* (Cambridge, Massachusetts, M.I.T. Press, 1971), Chapter 8, "The Motor Pump: A Subsistence Innovation," pp. 164-179 for documentation of the Vietnam case.

22. Chester I. Barnard, *The Functions of the Executive,* (Cambridge, Harvard University Press, 1938).

23. Herbert Simon, *Models of Man,* (New York, Wiley, 1957) pp. 198-9.

24. It is relevant to note that the Bank's report on "East Pakistan Tubewells Project," the basic document that top decision-makers received from the project appraisal mission, contained no discussion of alternatives to the type of technology recommended.

APPENDIX NOTES

1. F.A.O. *Monthly Bulletin of Agricultural Economics and Statistics,* No. 9, Vol. 19, (Rome, 1970) p. 46.

2. *Ibid.*, No. 4, Vol. 22, (Rome, 1973) p. 37.

ECONOMIC MAN AND ENGINEERING MAN:
Choice of Technology in a Low-Wage Country

Louis T. Wells, Jr.

Acknowledgments

The field research on which this paper is based was done while the author was a consultant to Harvard Development Advisory Service. Sjahruddin and Sri Hartono were of great help in the collection of the underlying data. I am grateful for the comments of Robert B. Stobaugh and Raymond Vernon on an earlier draft.

INTRODUCTION

One of the most perplexing problems facing a number of developing countries is that of unemployment. Industrialization, it is generally thought, has not done its share to absorb labor. It appears that the technology which is used early in the industrialization process is capital-intensive, borrowed from the advanced countries. In the opinion of many, a part of the answer to the unemployment problem is to turn to "intermediate technology," technology that involves less capital and more labor than that in general use in more advanced countries.

If the businessman behaves like an "economic man," he should, according to traditional wisdom, choose an intermediate or a labor-intensive technology when operating in a less developed country with low wage rates and high capital costs. By choosing appropriately, he will minimize his costs of production and simultaneously raise employment. Although much has been written about the possibilities, relatively little field work has been done.

Most of the literature on the choice of technology fits into one of four categories. Some work deals with what the criteria should be for a country faced with a choice of alternative technologies.[1] Some discusses the feasibility and costs of alternatives, in the form of studies of various possible techniques.[2] Other studies ask what can be learned of the choice of entrepreneurs from an examination of aggregated data.[3] Very little empirical work has been done at the firm level to examine how the businessman actually behaves.[4] This study of technology in Indonesia is an attempt to help close the gap.

The principal finding of the research is that the businessman probably does not behave solely as an "economic man." Economics puts a constraint on his behavior, but he may also have a rival objective function, which could be described as that of an "engineering man." Policies to encourage the use of intermediate technology must take into account this rival objective function, which appears to influence the design of plants in less developed countries.

THE RANGE OF TECHNOLOGY

The first question mark about the appropriateness of the model of the economic man comes when one is confronted with the wide range of technology that exists side by side in Indonesia. Some of the differences in technology can be explained by the fact that different investors face different factor costs, but a large unexplained residual is left over.

For this study, data were collected on the technology used in 43 industrial plants in Indonesia (Appendix A, p.87). Factories were visited in Djakarta, Bogor, Surabaya, Malang, and Medan. In six industries, plastic sandals, cigarettes, soft-drink bottling, bicycle and betjak tires,[5] flashlight batteries, and woven bags, data covered four or more plants producing comparable products.[6] For this core sample of plants, systematic comparisons of technology could be made. The remainder also appeared to fit the patterns observed in the core sample and reported in the tables. The quotes draw freely from the larger sample.

In four of the six industries, there seemed to be technologies that lent themselves to classification into "capital-intensive," "intermediate," and "labor-intensive" categories. Appendix B, p. 88, gives details of how the plants were classified; Table 1 gives the number of firms in each category by industry.

The difference in employment created under these alternative technologies is rather striking. To take the cigarette industry as an example, where calculations of the number of workers required to produce one million cigarettes per month are a common measure of automation, the employment implications are significant. In the capital-intensive class, the median of the range of number of workers required was 3; in the intermediate class, 6; in the labor-intensive class, 40. For the volume of cigarettes manufactured by the firms interviewed, this would mean 3,009 workers if all firms were using the capital-intensive technology, 6,018 if under intermediate, and 40,120 if under the labor-intensive techniques. Table 2, giving crude estimates of the number of workers required, shows that the choice of technology has similarly important implications for the employment in other industries.

TABLE 1

Number of Plants by Technology and Industry

Industry

Technology	*Plastic Sandals*	*Cigarettes*	*Soft-Drink Bottling*	*Bicycle and Betjak Tires*	*Flashlight Batteries*	*Woven Bags*	*Total*
Capital-intensive	2	3	1	1	2	2	11
Intermediate	6	5	3	4	2	2	22
Labor-intensive	0	3	2	1	4	0	10
	8	11	6	6	8	4	43

TABLE 2

Number of Workers Required in Various Industries for Various Technologies

	Industry					
Technology	*Cigarettes*	*Flashlight Batteries*	*Soft Drinks*	*Tires*	*Woven Bags*	*Plastic Sandals*
Capital-intensive	3,009	464	279	—	—	230
Intermediate	6,018	1,392	698	1,207	—	920
Labor-intensive	40,120	5,568	6,417	7,100	—	—

DIFFERENT FACTOR COSTS

The range of technology observed in Indonesia could of course be explained by market imperfections that lead different firms to have different factor costs. Each manager may be behaving as an economic man, arriving at solutions to the problem of choice of technology that are different from those of some of his competitors because of the different costs of labor and capital. The over-all data give some credence to this thesis. However, a more detailed examination sheds considerable doubt on the ability of different factor costs to explain the range of technologies adequately.

It is true that the foreigner is disproportionately represented in the firms using the more sophisticated technology.[7] And he is likely to face a lower cost of capital than the domestic firm. Table 3 clearly demonstrates the higher propensity of the foreign firm to adopt a capital-intensive process. The two foreign firms listed as using intermediate technology were interesting cases. One had plans to buy new equipment that would put it in the capital-intensive class. Both were employing secondhand equipment drawn from their factories in other countries.

Three further bits of evidence are available to support the idea that the breadth of technology might be a response of an economic man to different labor costs.

First, the foreign firms rather consistently pay higher wages than do the domestic firms, as Table 4 shows. The differences are large for both the skilled and the unskilled jobs and cannot be accounted for by differences in industry mix, location, or payment in kind. Thus, the foreign plant's choice of capital-intensive technology could result not only from access to cheaper capital, but also from the fact that the foreigner pays higher wages than the domestic enterprise.

The second bit of evidence comes from the subsidized "medium-term investment-credit program." Those firms that have received money at

TABLE 3

Number of Plants by Technolgy and Nationality of Investor *

Technology	*Nationality of Investor* *Foreign*	*Domestic*
Capital-intensive	8	3
Intermediate	2	20
Labor-intensive	0	10

* If any of the equity was held by a foreigner, the plant was supposed to be registered with the Foreign Investment Board. This registration was the basis of classification.

12% annual interest under this program seem generally to have chosen a more sophisticated technology than their domestic competitors who are forced to pay the market rate of 24 to 36%. One of the three domestic firms using capital-intensive technology was financed under this program, while only three of the 20 domestic firms using intermediate technology and none of the firms using labor-intensive processes were employing this subsidized capital. Additionally, the state enterprises seem to be consistently more automated than their private counterparts. One of the domestic enterprises employing a capital-intensive process and not financed under the "medium-term investment-credit program" was state-owned. Two of the plants employing intermediate technology were state enterprises, while none of those with labor-intensive technology were owned by the government.

The state enterprises may receive subsidized capital; Table 4 indicates that they face higher wage bills than their private competitors. Their choice of more sophisticated technology may be economically justified.

TABLE 4

Wages in Rupiahs* per Day, by Skill Level and Nationality of Investor

Skill Level **	*Nationality of Investor* *Foreign*	*Domestic [State]*	*Domestic [Private]*
Unskilled	314	227	123
Skilled	503	425	352

* One U.S. dollar equaled approximately 413 rupiahs at the time.
** Comparable jobs were taken in each industry. Industry by industry, the same patterns of wage differences were evident.

However, some of the statements of the managers of various enterprises raise questions about the usefulness of the simple model of the "economic man" for fully explaining the decisions that are being made.

The following quotes were not atypical:

1. After noting that machines in the process of being scrapped were capable of producing a high-quality product, one foreign manager explained the replacement of the intermediate equipment with sophisticated equipment as follows: "You have to modernize to stay ahead of the competition." This statement was made in spite of the facts that there was a ban on the import of the product and that there were no other firms manufacturing the product locally. When asked what he was going to do with the old equipment, he explained that he was going to cut it up and scrap it. He would not sell it because "some of these Indonesians can get any old equipment running."

2. The manager of a hand-rolled-cigarette factory answered the question as to which technology was cheaper by assuring me that automated cigarette factories had the advantage. When asked why he did not introduce machines, he explained that "the interest payments on the money we would have to borrow would exceed our wage bill." When challenged with the possibility that this statement might be inconsistent with his claims on costs, he responded that the automated plant would be cheaper in "the long run."

3. Another manufacturer who was employing young girls to attach labels by hand to his products explained that he was ordering a machine to replace them. Asked whether it was cheaper to attach them by machine, he explained that he did not know, but the girls were a lot of trouble: "They just cause management problems."

4. The manager of a plant with both an automatic and a semi-automatic line explained that he was converting the semi-automatic to fully automatic as soon as possible. He wanted to produce a "high-quality product." The output of both lines was already meeting the standards of the foreign licensor and was considered among the best in Indonesia by the firm's competitors. It was not clear that further automation would improve the quality, at least in a way that consumers would notice.

If the different factor costs are the principal considerations that lead to the different choices of technology, the engineering data should support the choices by indicating that the capital-intensive technology is suitable for the foreign firm and the domestic firms with subsidized capital. The data, however, point to a rather different conclusion. At any reasonable set of capital costs that a firm in Indonesia might face, the economics of the situation point to the conclusion that the choice of capital-intensive technology requires an investment per worker saved that is far beyond that which would be consistent with the actual wages paid by the domestic firm, the foreigner, or the state enterprise.

THE INVESTMENT TO REPLACE A WORKER

One way of determining whether firms are behaving as classical economic men is to calculate the incremental investment made per worker saved in one level of technology compared to the next lower level. The data collected from firms enabled calculations to be made for several of the industries.

In the choice between the labor-intensive processes and the intermediate level, the model of the economic man appears to be of some value in explaining the choice of technology. Unfortunately, only the cigarette and soft-drink bottling industries could be analyzed for this step.[8]

The incremental investment for machine rolling of cigarettes, compared to hand rolling, was between $50 and $120 per worker saved. For soft-drink bottling, the additional investment for semi-automatic bottling compared to hand bottling was on the order of $833 per worker saved.[9]

The wages paid out to cigarette rollers run on the order of $100 per year. Thus, even with the higher investment figure of $120 per worker, a capital cost of 36%, and a maintenance or depreciation cost of 10% per year, the annual capital cost of saving a worker is only about $55. These figures suggest that hand rolling of cigarettes continues only because of the unavailability of capital, even at 36% interest, because of a differentiated excise tax that favors hand-rolled cigarettes, or because the consumer prefers the "low-quality" hand-rolled version.

On the other hand, semi-automatic bottling may well be a poor investment for firms that must pay 36% for their capital. The annual capital and maintenance costs would be about $383, considerably above the wage rates in Indonesia. Note, though, that a foreign firm, with an 8% interest rate and the same 10% maintenance cost, would face an annual capital cost of only $150, less than the annual cost of an unskilled worker in a foreign-owned plant. It is thus perhaps not surprising that hand bottling was done by domestic producers only. Semi-automatic equipment was operated by both foreign and domestic-owned enterprises.

In the case of the choice between the intermediate technology and the capital-intensive level, the findings are rather different. Calculations could be made for critical processes in five of the six industries. The average investment per worker saved was $6,986. The range was from $2,000 to $21,500 (see Appendix C, p.91).

If the foreigner is able to raise capital abroad for 8%, and the replacement and maintenance requirements amount to 10% per year (or a 10-year life for the equipment), the annual wage that would have to be saved by an economic man would, for the average $6,986 investment, have to exceed $1,260 per year. The actual annual cash wage for employees of foreign firms averaged about $225. Even with a high cost for social expenses, it is unlikely that the real costs of a worker could approach $1,260 in Indonesia, nor is it likely that wages will reach that level in the foreseeable future.

At the low end, for an investment of $2,000 per worker the 18% capital costs amount to $360 per year. It is possible that the wages plus social costs of an employee for the foreign-owned plant could reach this level, but in only two industries was the investment below $3,500.

In spite of the fact that the capital costs involved in the capital-intensive technology generally far exceeded any possible wage savings, such investments were frequent and were being contemplated by many firms now using less automated techniques. The simple calculations suggest that these decisions did not make sense from the point of view of an economic man who wants to minimize costs.

OTHER ECONOMIC FACTORS

Three more complicated economic justifications should be disposed of before the model of the economic man is modified substantially or is allowed to be pushed to the back seat:[10] (1) The costs of developing more labor-intensive techniques may exceed the benefits of using known technology for the foreign firm; (2) the quality of output may not be sufficiently high or consistent with that of the intermediate technology; and (3) the intermediate technology may waste raw materials to an extent that their costs outweigh the benefits of the net savings on capital and labor.

These themes appear so frequently in the literature on choice of technology that they almost certainly have validity in many instances. They did not, however, seem to go very far in explaining what was observed in the light manufacturing plants included in this study.

Where the intermediate technology has to be developed from scratch, or even has to be copied from others, the development costs may well outweigh the benefits to be derived from lower operating costs. One of the plants visited while interviews were being conducted for this study had attempted to introduce a new technology. After large expenditures on engineering and construction, the manager found himself with machines that did not function. Further, unexpected expenditures on modifications were already under way. However, in general the managers of the plants included in this study of light manufacturing were well aware of and understood the intermediate alternative. In fact, many of the foreign firms had taken over a going enterprise as a joint venture, or had been given back a plant that had been nationalized under the Sukarno regime. In most such cases, the plant was already operating with an intermediate technology. However, in spite of the economics and knowledge of alternatives, the firms had already converted, or were in the process of converting, the plant to a more capital-intensive technology.[11] Ignorance or cost of developing intermediate technologies was not the problem.

The quality issue is a difficult one. Engineering standards and consumer standards of quality do not always seem to agree. Quality was frequently mentioned as a reason for increased automation. However, several facts point to the conclusion that the supposed quality differences were either not real or else were not perceived by the consumer as being real.

In one industry, product test data from engineering tests were available from three firms, covering the output of competitors. The products that ranked highest in level of quality and lowest in variance from this level came from manufacturers whose processes extended well into the range of intermediate technology. One of these plants, one mentioned in the previous quotes, was operating an automated line next to an intermediate technology line. The output of both was meeting the high standards of the foreign licensor. In other industries, I asked firms to rate the quality of competing products. Again, the highest rankings extended well into the range of intermediate technology.

In the case of cigarettes, a rather special industry in Indonesia, the engineering standard of quality seemed to conflict directly with the standards of a sector of the purchasers. Machine-rolled kretek cigarettes[12] were having a difficult time taking a share of the market from the hand-rolled ones. The even-burning, crackle-free characteristics of the machine-rolled seemed to be less liked by the kretek smoker than the uneven, noisy, hand-rolled variety. Machines had also not been developed to reproduce the slightly irregular, flared shape of the hand-rolled variety, which seemed to be more popular with the consumer.

The labor-intensive techniques did generally result in a product that was inferior, at least by engineering standards, to that of the more capital-intensive technologies. The hand-bottled soft drinks were, for example, probably not overly hygienic. Foreign matter could easily be introduced into the bottles by accident. And flies swarmed around the ingredients before they were hand-mixed. The flashlight batteries made by the labor-intensive techniques were more subject to leakage than were the others. The case of consumer preference for hand-rolled cigarettes is probably an anomaly. However, the intermediate technologies did seem capable of turning out products of a quality that was consistently as good as that of the capital-intensive techniques.

Contrary to expectations, the saving of raw material was never mentioned by the firms that were interviewed as a reason for choosing automated production processes; in fact, it appeared occasionally as a reason for not automating. The automatic looms in the bag-making plants produced more rejects than did the semi-automatic. Simple manual cutters got more sandal soles out of a piece of foam plastic than did the automatic, ganged ones. And so on. In process industries, saving of raw materials may be a justification for automation, but not in the firms in this sample.[13]

Even these somewhat more sophisticated versions of the model of the economic man seem insufficient to account for the choice of technology. One must look for other explanations.

OLIGOPOLY AND CAPITAL INTENSITY

One impression which was consistently reinforced in the interviews was that the Indonesian entrepreneur wanted as much as his foreign counterpart to have an automated plant. The suggestion was that somehow the foreign firm was in a position to fulfill this desire in more cases than was the domestic investor.

In fact, a better explanation than the nationality of firms that had chosen capital-intensive techniques seemed to lie in the degree of monopolistic advantage that the firm was able to exercise. The monopolistic position of such firms was evidenced by several factors. All but one of the foreign firms interviewed were using international trade names. The one exception had been operating in Indonesia for more than thirty years; it had developed brands that it did not use in other countries but which were thought by most consumers to be international.These brand names had given these foreign firms a position such that price was not the basis of competition. On the domestic side, monopoly also played a role. Of the three domestic firms using advanced technology, two had monopolistic positions. One was using an international trade name under license; the other was a state enterprise that sold almost all of its output to another state enterprise.[14] In the intermediate-technology category there were only two domestic firms using international trade names. Both of the plants were noticeably more automated than the other domestic firms in the same industry.

Price serves as another indicator of monopoly position. The

average price of the product of the capital-intensive plants was far higher than that of the intermediate level in all cases for which data were available (Table 5). Although the firms with capital-intensive technology all produce high-quality products, these products do not appear to rank higher in level of quality or consistency than those of the best of their competitors who do not have international trade names. Their prices, though, are higher.

A further hint that monopolistic advantage, not "foreign-ness," underlies capital intensity is given by the cases of foreign firms outside the industries studied systematically.[15] In these cases, my impression was that where trade names were not important, firms appeared to employ a much more labor-intensive technology than they used at home.

THE DRIVE TOWARD AUTOMATION

There appear to be a number of factors that drive firms toward more capital-intensive techniques than would be suggested by the simple economic models. In many cases the managers appear to want to respond to certain elements of risk and uncertainty. Capital-intensive plants seem, in some instances, to serve an insurance role. And some managers seem to be driven not only by simple economic objectives but also by objectives that could, perhaps, be best described as those of an "engineering man." But the ability of a manager to pay the cost of insurance against risk and uncertainty or to let the engineering objectives have their play is much greater for the firm that has a monopolistic position than for the one that faces severe price competition.

There are at least two kinds of insurance coverage that a capital-intensive plant might offer but the labor-intensive plant does not:

TABLE 5

Approximate Average Retail Price of Product, in Rupiahs

	Industry				
Technology	*Plastic Sandals* [1]	*Cigarettes* [2]	*Soft Drinks* [3]	*Bicycle Tires* [4]	*Flashlight Batteries* [5]
Capital-intensive	120	78	63	470	55
Intermediate	60	60	36	350	40

[1] One-piece sandals, per pair.
[2] The most popular brand, for each plant, per package of 20, white cigarettes only.
[3] For the contents of a bottle, adjusted to reflect volume.
[4] Wholesale.
[5] Size "D" (UM-1).

the capital-intensive plant may allow the manager to respond more quickly to unexpected fluctuations in demand or to levels that are different from that for which the plant was designed; and the capital-intensive plant may be perceived as allowing the manager to reduce the risks of facing future liquidity crises.

Consider the problem of an entrepreneur who confronts a market for which the level of sales is uncertain or is likely to vary considerably. If the technology that minimizes costs is one that makes it difficult to adjust plant output to levels other than the optimal design output, the entrepreneur may decide not to adopt it; he may instead adopt a technology that allows for easy adjustment of output, even though it may entail a higher unit cost of production.

In many cases, the more capital-intensive techniques are more readily adjusted to different levels of demand. Confronted with sporadic and unpredicted increases in demand, the manager can more easily run an added shift in an automated plant than he can install the necessary complement of workers in a labor-intensive plant. In the labor-intensive plant, large numbers of additional workers will have to be recruited and trained. With brief contractions of demand, automated plants can be adjusted by running the machines more slowly. In labor-intensive plants, the alternatives of laying off workers or slowing the work pace are particularly unattractive. Layoffs are expensive or difficult in a country such as Indonesia, where the labor laws discourage the practice. And slowdowns may result in work habits that are hard to change when production is to be expanded again. Of course, small fluctuations in demand may be handled out of an inventory manufactured at constant output levels. However, in many cases, the storage costs and the risks of product obsolescence or deterioration will severely limit the ability of inventory to match a constant output to a fluctuating demand.

If the insurance against errors in demand projections is attractive, perhaps the firms facing price competition should behave as the monopolists do. However, an entrepreneur faced with keen price competition is worried that his competitors may not saddle themselves with the insurance premiums in the form of higher production costs. In that case, the insured firm faces a lopsided loss-gain situation. If the original market forecasts turn out to be correct, the insured firm may be driven out of business by those not having to bear the insurance costs. At worst, the uninsured firm may miss some opportunities to profit from a quick expansion of its output. Its response, therefore, may be not to pay the insurance but to train a work force capable of producing at a level on the low side of its projected sales.

On the other hand, the firm with a monopoly rent that derives from its inelastic demand curve may be able to afford the insurance. It can bear the higher production costs at the expected demand level and supply a larger market or contract easily if the market turns out to be above or below that for which the plant was designed to be optimal.

When the monopolistic firm is vertically integrated, the gains in being able to respond to an unanticipated increase in demand may be larger than is indicated by the profits of the operation under consideration. If the operation is only an assembly, bottling, or packaging plant, the increase in demand opens up the possibility of selling added inputs from existing capacity in other plants. The output of such plants will, as a rule, face declining costs. The design of a foreign-owned pharmaceutical plant provides a good example. The plant makes tablets and capsules from the products of factories located elsewhere but owned by the same parent. The output of this capital-intensive plant can be expanded easily to meet unanticipated demand. The incremental contribution to the multinational system of plants from the sale of a bit more chemicals is very high. The higher costs of the capital-intensive plant at the final stage can easily be offset by the additional profits that derive from the use of more intermediates from affiliated firms whenever small increases in demand are encountered.

The second kind of insurance that a more capital-intensive plant may offer is that the chances of a liquidity crisis will be reduced. Managers responding to this kind of issue claimed that capital-intensive plants were "cheaper in the long run" in spite of the fact that an economist would argue that the labor-intensive operation had lower costs. The managers' behavior is at least partly a result of the imperfections in capital markets.

Consider the possibility that a firm, confronted with monopolistic profits for a few years, fears the possibility that price competition may become so severe in the future that prices will fall close to marginal costs. Prices may actually fall to the point where they approach the marginal costs of a capital-intensive manufacturer rather than the higher marginal costs of a labor-intensive firm. The entrepreneur in Indonesia is constantly reminded of this possibility by some plants that had been closed as a result of intense price competition.

Faced with this possibility, the businessman could behave in one of at least two ways. He could build a labor-intensive plant with minimum costs and invest the incremental cash resulting from the savings during the period of relatively high prices in assets that would yield a return. If prices fell below his marginal costs later, he could close the plant. He would still have an income from the assets that he obtained as a result of his lower costs during the period of high prices. This income would presumably be as great as the cash flow that the competition with a capital-intensive plant would be generating out of staying in business. But two factors make this strategy unattractive. First, the businessman perceives a high risk attached to alternative investment for his assets in imperfect capital markets, and he is reluctant to withdraw from his basic business or to remain in a business with a cash deficit that would have to be subsidized.

An alternative strategy would be to borrow to build a more capital-intensive plant at the beginning, pay off the loan during the period of high profits, and stay in the same business during the competitive period as long as prices remain above his low marginal costs. This strategy generally appears much more attractive than that of minimizing costs. If intense price competition becomes a reality, the entrepreneur can stay in the same business without having to make up a cash deficit. And he can avoid the risk of having to invest assests in the capital market to cover later losses.

Although the strategy that minimizes cash costs in the future appeared to be more attractive to most Indonesian businessmen, the only ones who could follow it were generally those who held a sufficiently monopolistic position so that they could bear the heavier burden of financing for the capital-intensive plant until a loan could be repaid.

While part of the drive toward capital-intensive technology may be explained by the desire of the oligopolist to insure against risk and uncertainty, a large part seems also to be a response to some objectives of the engineering man. These are:

1. Reducing operational problems to those of managing machines rather than people.
2. Producing the highest quality possible.
3. Using sophisticated machinery that is attractive to the engineer's "aesthetics."

Of course, reducing management problems could be viewed as an economic objective if management is considered as a production factor and management attention that is freed from problems associated with people could be devoted to other money-making activities, with a higher return. However, there is no reason to expect the opportunity cost of managers of plants in an oligopolistic position to be higher than those of plants in a more competitive market. The drive appears to be only partly a response to economic opportunity costs.

Similarly, the production of a high-quality product could be viewed as an economic objective so long as quality increases the sales sufficiently to offset the higher production costs. But the concept of quality appeared to go far beyond that which had a meaning to the customer. And, although variances in quality would be a major problem in the very labor-intensive technologies, it did not seem to be an issue with the intermediate techniques.

It should not come as a surprise that engineering objectives play a role in the choice of technology. After all, most plants are designed by engineers. And that engineering objectives should sometimes depart from economic objectives should also not be surprising. It has long been recognized that scientists are responsive to objectives not easily captured in economic models. Engineers also have a training that may instill in them noneconomic values. Responsiveness to these values is no less "rational" than is a response to economic values.

It is worth noting that few of the Indonesian firms did any calculations to investigate the appropriate technology to employ. On the other hand, most of the foreign firms required some kind of feasibility calculation before new equipment was installed. However, such studies did not generally compare one technology to an alternative. The most common type of calculation was an engineering study in which the economic return was calculated only for the equipment proposed by the engineers; no comparison was made to alternative techniques. In fact, in one case the calculations were done in the United States by a feasibility team made up of engineers who had never visited Indonesia. The suggestion was that U.S. wage rates might have been used in the calculations. The manager explained, when asked about the effect of lower wage rates, that "you get what you pay for in productivity."

CONSTRAINTS ON AUTOMATION

If a firm is forced to compete on a price basis, economics provides a brake on the degree to which management can allow the engineering objectives to override the economic ones and a limit on the ability of a firm to use capital-intensive technology to insure against risk and uncertainty.[16] Our research indicated that there are other brakes as well. In fact, only one of the sophisticated plants was described by management as being as automated as would be a similar plant in an advanced country.

The principal restraint for the firms with inelastic demand seemed to be the problem of maintaining and operating advanced equipment in a developing country. In fact, one domestic firm had installed very sophisticated machinery, but was gradually disconnecting the automatic controls as they broke down. And most of the complaints about the quality of Indonesian labor came out of the plants with the most capital-intensive technology. Maintenance and operation of the sophisticated equipment were requiring scarce skills.

The labor laws appeared to provide some brake for firms already in operation. A few managers explained that they had firm plans to automate, but could not do so immediately because they were not allowed by the government to release the workers made redundant. As soon as natural attrition of workers enabled them to do so, they would proceed to a more sophisticated technology.

And for some of the very labor-intensive firms, the inability to raise capital at any price seemed to act as a brake on moving to intermediate technology. Some were unable to borrow from the state banks and did not have the contacts to raise money on the quasi-legal private money market. What from the private economic view looked like an attractive movement up the scale of technology was constrained by the unavailability at almost any price of the required capital.

CONCLUSIONS

This study has only begun to probe the complex factors that influence the manager in his choice of technologies. The simple combination of production functions and factor costs is clearly inadequate to the task. There are many promising hypotheses waiting to be tested. A surprisingly wide range of technology was found to exist within the same industry and country. The breadth is probably partly a result of the fact that the study examined only light manufacturing. However, the factors that influence the managers in their choice of technology are probably similar to those in other kinds of industries.

The capital-intensive techniques were generally associated with foreign investors. However, they seem to be more closely associated with investors who had some monopolistic position. Foreigners happened to be in such a position more frequently than domestic investors. But the drive toward capital-intensive plants seemed to be similar among both foreign and domestic managers.

The managers' choice of technology appears to be influenced by two objective functions, which, in low-wage countries, are generally conflicting. The first objective, that of the "economic man," is to minimize costs. This leads to a relatively labor-intensive production process. On the other hand, the objective of the "engineering man" tends to lead toward more sophisticated, automated technology. Where price competition is the rule, the objectives of the economic man seem to override those of the engineering man. However, when the firm has a monopolistic advantage, there is less pressure on the firm to minimize costs to survive; the goals of the engineering man are allowed to move the firm to a level of technology that is more advanced than that which the economic man would choose.

Although the foreign firms are frequently in the position of having an oligopolistic position, local firms appear in this study to behave the same as the foreign firms when they have such an advantage. Both have their engineering men.

Additionally, capital-intensive plants appear to provide some degree of insurance against risk and uncertainty. Under some circumstances, only the firms that have a monopolistic position can afford to pay the insurance premium, in the form of higher production costs, associated with the capital-intensive choice.

The relative importance of the insurance factors and the engineering objectives have not yet been examined.

In more advanced countries there may be little conflict between the economic and the engineering objectives. With high labor costs, automation is the direction sought under both objectives, and insurance is gained automatically. It is possible that the range of technologies existing side by side in an advanced country will be less than that found in Indonesia.

Considerable evidence has been accumulated in macrostudies; it shows that the average capital-to-labor ratio for individual industries in the high-wage countries is greater than that in the low-wage countries. It is possible that the range of technology employed in the latter is also greater than that employed in the high-wage ones.

Although relative factor costs play a role in the selection of technology, there appear to be other influences on the managers' decisions which are of critical importance to the policy maker interested in the employment effects of industrialization.

Appendix A

AGE AND SCALE OF PLANTS

Two of the difficulties encountered in cross-sectional comparisons of technology were minimal in this study: the inclusion of old plants along with new ones, and the problem of economies of scale.

Of the 43 plants studied, 20 went into production in 1965 or later. Of the 23 plants that were established earlier, 9 had made substantial additions to their equipment since 1965. The more recently introduced technology was used as the basis of classification. Six of the remaining old plants were in the cigarette industry. Although the foreign plants accounted for the same percentage of old plants as they did of the whole core sample, none were represented in the plants that had no major additions since 1965.

There was considerable overlap in the size of plants in the capital-intensive and intermediate categories. In all but one of the industries, there were plants using intermediate technology that had an output at least as great as one of the plants using capital-intensive technology. And in three of the industries, the capital-intensive plants and the intermediate plants had several machines at the principal stage running side by side. However, the labor-intensive plants were generally considerably smaller than their counterparts with a more advanced technology.

Appendix B

THE METHOD OF CLASSIFICATION

Classification of technology can be done in a number of ways. Most popular, in studies at the macro level, is the use of capital/labor ratios. In this study an attempt to use capital/labor ratios at the firm level was quickly discarded, because of several difficulties. The principal one was the problem of finding an appropriate measure of the value of the capital employed. The bookkeeping of the firms interviewed was not consistent. Different rates of depreciation were used. Purchase at different times led to different values for undepreciated equipment. Book value sometimes seemed to reflect tax and duty considerations, especially when equipment was purchased through a foreign affiliate. And books kept in rupiahs were useless where the life of the firm extended back into the period of rapid inflation in Indonesia. Moreover, firms were hesitant to show their books. The second problem came from the fact that the degree of vertical integration was not the same from firm to firm. Although the number of workers could be adjusted, approximately, to reflect a common level of integration, the capital figures were terribly difficult to disaggregate. An additional difficulty lay in accounting for the "inferior technology" found in a few cases. In these instances, both the capital and the labor required to produce a given output were greater than that under some alternative technology. Although such technology was not encountered in many cases, a mechanistic measure might have distorted the findings.

As a result of these difficulties, plans were classified principally by the kind of equipment in use. (Where new equipment was actually on order, it was counted as being in use.) Some key processes were identified in each industry that seemed to be good reflections of the level of automation throughout the plant. These classifications were checked, to the extent possible, by using labor/output ratios, where labor was adjusted to comparable levels of vertical integration. The labor/output ratios were crude, but they enabled some approximations to be made of the employment implications of the various levels of technology.

The categories of technology ("capital-intensive," "intermediate," and "labor-intensive") are obviously rather arbitrary. To protect the confidentiality assured the firms, some way had to be found to aggregate the industries. The author hopes that the reader will take it on faith that the disaggregated data contained no surprises.

The classifications were based on the following criteria:

Cigarettes

Capital-intensive: Use of 2000-per-minute cigarette makers, and machines for all stages of tobacco preparation. Average of 3 workers per million cigarettes per month.

Intermediate: Use of 1300-per-minute or slower cigarette makers. With one exception, some hand preparation of tobacco. Average of 6 workers per million cigarettes per month.

Labor-intensive: Hand rolling of cigarettes and primarily hand preparation of tobacco. Average of 40 workers per million cigarettes per month.

Flashlight Batteries [17]

Capital-intensive: Linked conveyor system, most of the parts assembled by machine. Average of 1 worker per thousand batteries per day.

Intermediate: Unlinked conveyor system, some parts assembled by machine. Average of 1 worker per thousand batteries per day.

Labor-intensive: Unlinked conveyor system or no conveyors, all assembly by hand. Average of 12 workers per thousand batteries per day.

Soft Drinks

Capital-intensive: Automatic uncrating of bottles, loading of bottle washer, and crating of bottles. Average of 1 worker per thousand bottles per day.

Intermediate: Bottling machine, but crates unloaded and loaded by hand. Average of 2.5 workers per thousand bottles per day. There was a considerable difference between the top and bottom ends of this category.

Labor-intensive: Hand washing and hand filling of bottles. Average of 23 workers per thousand bottles per day.

Tires

Capital-intensive: Use of Banbury mixer, automatically timed curing molds, and completely power-driven calenders. Automatic bead-making and bias-cutting equipment. Unable to estimate average number of workers.

Intermediate: Use of millers, partially hand-driven calenders. Untimed curing molds. Semi-automatic or hand making of beads and bias cutting. Average of 8.5 workers per 100 tires per day.

Labor-intensive: Most operations by hand. Average of 50 workers per 100 tires per day.

Woven Bags

Capital-intensive: Use of 200 pick-per-minute automatic looms, 1 operator per 4-6 looms.

Intermediate: Use of 150 pick-per-minute or slower looms, 1 operator per loom. Over-all workers per unit output could not be calculated, due to the use of different fibers, requiring different preparation.

Plastic Sandals

Capital-intensive: Use of automatic injectors, requiring 1 to 1.5 operators per injector. Approximately 5 workers per thousand sandals per day for simple sandals.

Intermediate: Semi-automatic injectors, with 2-5 operators each. Approximately 20 workers per thousand sandals per day for simple sandals.

Appendix C

THE INVESTMENT PER WORKER

For the investment per worker saved by employing the intermediate technology instead of the labor-intensive technology, the processes described below were compared. In each case, the cost data were collected from firms that were using the technology, and the number of machines and workers was adjusted to reflect a given output level.

1. Cigarettes: Use of secondhand 1300-per-minute makers or 2000-per-minute makers, modified to handle clove mix, in comparison to hand rolling, with a simple wooden roller. Secondhand machines appeared to be readily available, from abroad or from local plants that were automating.
2. Flashlight batteries: Insufficient data available.
3. Tires: Insufficient data available.
4. Soft-drink bottling: Use of a simple semi-automatic bottling machine in comparison to hand filling of bottles.

For the incremental investment per worker saved by employing the capital-intensive technology instead of the intermediate, the following processes were compared:

1. Cigarettes: The use of new 2000-per-minute makers in comparison to the use of second-hand 1300-per-minute machines.
2. Plastic sandals: The use of automatic injectors in comparison to the use of new semi-automatic injectors from Hong Kong.
3. Flashlight batteries: The use of an automatic conveyor assembly system from Japan in comparison to the use of a semi-automatic system from Japan.
4. Woven bags: The use of fully automatic looms in comparison with the use of semi-automatic flat looms.
5. Soft-drink bottling: The use of automatic crating equipment in comparison with the use of manual crating.
6. Tires: Insufficient data available.

Notes

1. See, for example, Alfred E. Kahn, "Investment Criteria in Development," *Quarterly Journal of Economics*, Vol. LXV, Feb. 1951; Hollis B. Chenery, "The Application of Investment Criteria," *ibid.*, Vol. LXVII, Feb. 1953; Walter Galenson and Harvey Leibenstein, "Investment Criteria, Productivity and Economic Development," *ibid.*, Vol. LXIX, Aug. 1955; R. S. Eckaus, "The Factor-Proportions Problem in Underdeveloped Areas," *American Economic Review*, Sept. 1955; Otto Eckstein, "Investment Criteria for Economic Development and the Theory of Intertemporal Welfare Economics," *Quarterly Journal of Economics*, Vol.LXXI, Feb. 1957; Amortya K. Sen, "Some Notes on the Choice of Capital-Intensity" in *Factors in Production* (Amsterdam: North-Holland Publishing Co., 1964); United Nations, *Industrialization and Productivity*, Bulletin 3, Department of Economic and Social Affairs, New York, March 1960; A. K. Sen, *Choice of Techniques* (Oxford: Blackwells, 1960).

2. See, for example, Gerard K. Boon, *Economic Choice of Human and Physical* Development Planning," *ibid.*, Vol. LXXI, Nov. 1957.

3. See, for example, Lloyd Reynolds, *Wages, Productivity and Industrialization in Puerto Rico* (Homewood, Ill.: Richard O. Irwin, Inc., 1965) Marc Nerlove, "Recent Empirical Studies of the CES and Related Production Functions," in *The Theory and Empirical Analysis of Production*, Studies in Income and Wealth, Vol. 21, N.E.B.R., New York, 1967; Il Sakong, *Factor Market Price Distributions and Choice of Production Techniques in Developing Countries*, unpublished Ph.D. dissertation, Grad. Sch. of Bus. Adm. UCLA, 1969; Zvi Grilliches, "Capital-Skill Complementary," *Review of Economics and Statistics*, Nov. 1969; R. Hal Mason and Il Sakong, "Level of Economic Development and Capital-Labor Ratios in Manufacturing," *ibid.*, Vol. LIII, No. 2, May 1971. For a summary of such studies, see Henry J. Bruton, "The Elasticity of Substitution in Developing Countries," Research Memorandum No. 45, Center for Development Economics, Williams College, Williamstown, Mass., April 1972.

4. For a particularly useful study in this classification, see W. Paul Strassmann, *Technological Change and Economic Development: The Manufacturing Experience of Mexico and Puerto Rico* (Ithaca: Cornell University Press, 1968). See also R. Hal Mason, "The Transfer of Technology and the Factor Proportions Problem: The Philippines and Mexico," *UNITAR Research Report No. 10*, United Nations Institute for Training and Research, New York, no date.

5. Betjaks are the pedicabs in common use in Indonesia. Betjak tires, although the same sizes as bicycle tires, are generally made for much heavier use.

6. If two distinct technologies were used by the same firm for a product, or if two products included in this study were manufactured by a firm, each product-technology combination was counted as a plant. Two processes for the same product in the same firm were encountered in three cases.

7. This is not a result of uneven distribution of foreign investors across industries.

8. Appendix C describes the processes analyzed to arrive at the investment per worker saved. The range of cost estimates for alternative machines varied greatly from firm to firm, indicating that alternative technologies had rarely been looked into in much detail.

9. All calculations were made for a constant output. For example, if to produce 1000 items per day two machines costing $500 and requiring two workers each were required at the labor-intensive level, and one machine costing $1,500 and using two workers were required at the intermediate level, the investment per worker saved would be $250.

10. The problems of different size plants and different periods of choice of equipment were relatively minor in this study. See Appendix C.

11. Note that they were replacing intermediate equipment, for which most of the costs would not be recovered. Thus, their investment in saving a worker would generally be higher than those calculated in the previous section.

12. Kretek cigarettes contain cloves, in contrast to the so-called "white" cigarettes.

13. A similar result was found for plywood manufacture in Korea. See Gustav Ranis, "The Role of the Industrial Sector in Korea's Transition to Economic Maturity," Center Discussion Paper No. 25, Economic Growth Center (New Haven: Yale University, October 1971). In fact, economic historians have claimed that the substitution of capital for labor in the industrialization of Western countries resulted in higher raw-material usage.

14. The third firm was in an industry that was protected and for which domestic capactiy was far below domestic demand.

15. In a perceptive, unpublished study, Wayne Yeoman pointed out that a U.S. firm was more likely to depart from its U.S. technology in low-wage countries if the primary basis of competition was price. See W. A. Yeoman, "Selection of Production Processes for the Manufacturing Subsidiaries of U.S. Based Multinational Corporations," unpublished D.B.A. thesis, Boston, Mass,: Harvard Business School, April 1968.

16. The Yeoman study, referred to in footnote 15, indicated that American firms adjust their technology more, in low-wage countries, for light manufacturing than for heavy manufacturing. The greater adjustment for light manufacturing probably reflects the existence of a wider range of known, feasible technologies for light industries.

17. With the many steps involved in the making of batteries and the several alternative processes, battery manufacture provided more of a continuum of technological alternatives than did the other industries.

IMPORT SUBSTITUTION, EMPLOYMENT, AND FOREIGN EXCHANGE IN COLUMBIA: NO CHEERS FOR PETROCHEMICALS

David Morawetz

Acknowledgments

I wish to thank Ana Rita Cardenas, Alejandro Figueroa, Gonzalo Giraldo, Mauricio Gutierrez, Thomas Hutcheson, Juan Klein, Fernando Sandoval, and Alberto Villate for helpful discussions and for assistance in data-gathering; and Carlos Díaz-Alejandro, Joseph Stern, Lawrence White, and an anonymous referee for comments on an earlier draft.

Like several other Latin American countries, Colombia began to develop her own petrochemicals industry in the 1960s. This development was a continuation of the import substitution policies begun after the Great Depression. The first phase of import substitution, production of "light" consumer goods, was more or less completed by the 1950s. By the mid-1960s domestic production had been substituted for imports in many "middle range" products as well (paper and paper products, some chemicals, and metal manufactures). This left only heavier, more capital-intensive industries, including petrochemicals, as candidates for further import substitution. Over 100 million dollars were invested in petrochemicals in Colombia in the 1960s, and another 120 million dollars are scheduled to be invested during the 1970s.[1] Comparable sums were also slated to be invested in petrochemicals in the other countries of the Andean Group—Chile, Peru, Ecuador, and Bolivia. In Colombia, petrochemicals received more government finance and support than any other two-digit industrial sector in the 1960s.[2] For example, in 1969 chemicals and petrochemicals represented 40 percent of the portfolio of the largest government industrial development agency (Instituto de Fomento Industrial) and received 25 percent of all credits and refinancings granted by the government's Private Investment Fund (Fondo de Inversiones Privadas), in spite of the fact that they generated less than 10 percent of industrial production and employment.

This paper analyzes the Colombian petrochemicals industry from an economic viewpoint, comparing past and projected future petrochemicals investments with potential alternative uses of the same funds in labor-intensive export-oriented industries—clothing, footwear, wooden furniture, textiles, glass products, and some metal manufactures. It is assumed throughout the paper that capital is a scarce factor of production in Colombia. Obviously, if capital were abundant the conclusions might well be different. The best way of comparing petrochemicals and labor-intensive investments would be to do a full social cost-benefit analysis, but that is ruled out by data limitations in the present case. Instead, several partial criteria are used to make the comparison: direct and indirect employment created, foreign exchange saved and earned, degree of international competitiveness of costs and prices, and linkage effects on the rest of the economy. None of these criteria on its own is sufficient to evaluate investment alternatives fully. However, if a project is found to be inferior on all four criteria, a strong prima facie case can be made for its rejection. In the present comparison, investments in petrochemicals are found to be inferior to labor-intensive export-oriented projects on all four counts.

THE PETROCHEMICALS INDUSTRY[3]

Petrochemicals is an extremely capital-intensive business. Only a handful of workers are required after a petrochemicals plant has been constructed, mainly to watch dials, test pressures, or do maintenance work. The only unskilled laborers involved are the cleaning staff.

Economies of scale in petrochemicals, which use a continuous flow process, are far greater than in car or steel making which, since their products are not liquids or gases, are confined to batch production methods. In petrochemicals, when the size of equipment is doubled its cost does not double; rather it rises by the smaller amount indicated by the following basic type of equation:

$$C = K\,(A/B)^b$$

The unknown cost (C) of an equipment of size A is equal to the known cost (K) of an equipment of size B multiplied by the ratio of the sizes A/B raised to the"b" power. The average value of the exponent "b" is about 0.6. By the "rule of 0.6" it has become virtually accepted in the petrochemicals industry that a big plant double the size of a small one will cost only 50 percent more to build, and (by the "rule of 0.2") will use only 15 percent more labor. Hence, the cost of capital per unit of production in a large plant is only 75 percent of the unit cost of capital in a plant half the size, and the labor cost per unit of output is only about 60 percent of that in the smaller plant. If the project under consideration is half a million tons of ethylene, such considerations may well dominate the planning process.

The history of the world petrochemicals industry over the last two decades has been one of capacity leapfrogging over demand, with a consequent general decline in product prices. The price of ethylene in Europe fell by half between 1966 and 1970. Thirty of the most important petrochemicals products in the United States fell in price by 40 to 60 percent in the decade to 1969. In the face of the enormous amounts of capital needed to construct a single economically-sized complex, these price declines caused several important international companies to have second thoughts about further expansion in petrochemicals in the early 1970s.

A study of the petrochemicals industry by M.A. Adelman and M.B. Zimmerman (1974) concludes that prospects for petrochemicals investments by developing countries which are not members of the Organization of Petroleum Exporting Countries (OPEC) are dim indeed.

Most non-OPEC developing countries are at a severe and continuing disadvantage in the industry because of lack of know-how, and the high opportunity cost of capital and feedstocks (raw or intermediate materials used in manufacturing chemicals). Further, the huge capital resources available to other countries, particularly OPEC members, and the low opportunity cost of feedstocks to OPEC countries (the opportunity cost for non-OPEC members is the export price, but for OPEC members, who restrict their export sales, it is the long-run marginal cost of production) suggest that larger and larger plants will continue to be built, driving prices still lower. In short, for most developing countries, "the best policy is non-entry."

The Colombian petrochemicals industry consists of 20 or so establishments, producing products including aromatics, fertilizers, mineral oils, synthetic fibers, and intermediate and final plastic products. Almost all the plants are relatively small by world standards, the largest costing less than $20 million to construct. Tariffs averaging 33 percent and, more importantly, strict import licensing and prohibitions, afford significant protection to domestic producers.

ECONOMIC EVALUATION

Like all Latin American countries, Colombia suffers from severe unemployment. In the late 1960s, there were more than twice as many people unemployed (700,000) as there were employed (300,000) in the entire manufacturing industry (Table 1). Furthermore, at least as far as manufacturing employment was concerned, the situation was deteriorating as time went on. Employment in manufacturing industry grew at less than 0.3 percent *per annum* from 1962 to 1969, far behind population growth of over 3 percent *per annum*. Heavy investments in petrochemicals in the 1960s did nothing to ease the Colombian unemployment problem; on the contrary, they aggravated it. Despite investments of over 100 million dollars in "derivatives of petroleum and coal" (petrochemicals and oil refining) between 1957 and 1965, total employment in the sector actually *decreased* by 290 persons in the same period.[4] In at least one petrochemicals plant owned and operated by the government, each job created required $500,000 of fixed investment. On the average, it took $39,000 of fixed investment to create a job in petrochemicals, twenty times more than in labor-intensive industry (Table 2). In other words, for each 100 million dollars invested in petrochemicals in Colombia, some 2,500 persons are employed, while investment of the same 100 million in labor-intensive industries would create 50,000 jobs. The direct unemployment created as a result of investing 100 million dollars in petrochemicals is therefore about 47,500 persons, or more than 15 percent of total Colombian industrial employment.[5]

TABLE 1
Industrial Employment in Colombia by Sector, 1956 and 1967

Sector	1956 Persons	1956 Percentage of total employment	1967 Persons	1967 Percentage of total employment	Ratio employment 1967/1956
Food	34,399	16	44,057	15	1.28
Beverages	11,936	6	15,362	5	1.29
Tobacco	5,450	2	3,469	1	0.64
Textiles	36,498	17	46,580	16	1.28
Clothing and footwear	29,116	14	29,037	9	1.00
Wood products	5,865	3	6,366	2	1.09
Furniture of wood	4,856	2	5,177	2	1.07
Paper products	2,968	1	6,183	2	2.08
Printing	8,711	4	11,884	4	1.36
Leather products	4,295	2	4,377	1	1.02
Rubber products	4,781	2	6,781	2	1.42
Chemicals	11,148	5	23,832	8	2.14
Derivatives of petroleum and coal	2,027	1	2,061	1	1.02
Non-metallic minerals	18,968	9	24,739	8	1.30
Basic metals	4,907	2	4,713	1	0.96
Metal products	8,216	4	20,224	7	2.46
Non-electric machinery	2,215	1	5,937	2	2.68
Electrical machinery	3,062	1	9,507	3	3.10
Transport	8,915	4	13,942	5	1.56
Other industries	3,645	2	9,597	3	2.63
TOTAL	211,978	100	293,825	100	1.39
Total unemployment			721,000		

Source: D.N.P., *Breve Esquema sobre el Problema del Desempleo en Colombia,* (DNP-590-UPG, June, 1970), Table 1.

Proponents of petrochemicals have argued that the petrochemicals industry creates more indirect employment than clothing, footwear, and wooden furniture. This is not the case at all. An increase in clothing production stimulates the production of textiles and raw cotton, both relatively labor-intensive, and also of sewing machines and looms. An increase in shoe production stimulates the leather, cattle and shoe-machinery industries. On the other hand, an increase in petrochemicals output stimulates oil refining and crude oil production, neither of which uses much labor. We can use the Colombian input-output table to quantify approximately the number of indirect jobs created by a given investment in a particular sector. Appropriate manipulation of the inverse of the input-output table indicates that an investment of 100 million dollars in "derivatives of petroleum and coal" creates fewer than 1,000 jobs indirectly throughout the economy, compared with 102,000, 56,000, and 49,000 indirect jobs created by an investment of the same magnitude in clothing, footwear, and wooden furniture, respectively.[6]

Furthermore, small-scale high-cost protected production of petrochemicals like that in Colombia penalizes the industries which use petrochemicals products as inputs—textiles, plastics, and so forth. To the extent that high protection is granted to user industries to enable them to absorb increased raw material costs, high-cost production of petrochemicals contributes to raising the general cost structure of the national manufacturing industry. But where the user industry is a potential exporter, actual *indirect unemployment* may be created as well.

Take the case of the Colombian textile industry. Despite the fact that synthetic textiles are subject to much less stringent import quotas in the developed countries than cotton textiles, and that synthetic textiles are sold at higher prices per square yard than cotton textiles, Colombia exported almost solely cotton textiles in the late 1960s and early 1970s—about 20 million dollars worth per year. The chief cause of this anomalous situation was that the domestically produced synthetic fibers which textile producers were forced to purchase were overpriced by up to 120 percent.[7] This in turn was partly due to excessive protection granted to the (subsidiaries of foreign) firms producing synthetic fibers, and partly to the high cost of Colombian petrochemicals products used to manufacture these fibers. Assuming that Colombia could export, say, 100 million dollars worth of synthetic textiles *per annum*,[8] employment in the Colombian textile industry could be increased by some 14,000 persons (almost five percent of total Colombian industrial employment) if textile producers were permitted to buy their raw materials at world prices.

The picture is no more favorable for petrochemicals if we look at foreign exchange earned and saved by domestic manufacture. In Colombia in the 1950s and '60s, foreign exchange earned and saved per unit of capital invested was on average five times as great in labor-intensive industries as in petrochemicals (Table 2). It is true that exports as a percentage of total sales were higher in petrochemicals than in labor-intensive industry. However, it is not legitimate to look only at foreign exchange earned without looking also at foreign exchange saved. The apparently successful export performance of petrochemicals mainly reflects the fact that once the tanks, pipes, and cauldrons have been erected, the *marginal* cost of producing petrochemicals is very low. Once a plant has been built, it is of course profitable to export as long as the ("dumping") price is greater than marginal cost. Furthermore,

TABLE 2

Capital-Intensive and Labor-Intensive Industry in Colombia

Industry	Fixed capital (US$'000)	Total employment (persons)	Fixed capital per job created = column 2:3 (US$'000)	Exports (US$'000)	Foreign exchange earned and saved (= total production)[a] (US$'000)	Exports/Production (per cent) = column 5:6 x 100	Foreign exchange earned and saved/Capital = column 6:2	Percent of capital that is national	Domestic price/FOB export price
1	2	3	4	5	6	7	8	9	10
Petrochemicals	5,680.0	146	39.0	263.0	3,486.0	7.5	0.6	55.2	1.47
Clothing	358.6	192	1.9	14.9	1062.7	1.4	3.0	99.1	1.16
Footwear	128.1	175	0.7	1.1	785.7	0.1	6.1	99.1	0.83
Furniture of wood	330.0	140	2.3	-	464.0	-	1.4	100.0	1.06
Textiles	5,158.8	1,731	3.0	765.2	12,757.3	6.0	2.5	100.0	1.32[b]
Glass products	435.8	252	1.7	-	368.3	-	0.9	100.0	0.98
Metal manufactures	339.8	142	2.4	26.4	1,282.4	2.1	3.8	89.5	1.30
Unweighted average of six labor-intensive industries	1,125.2	439	2.0	134.6	2,786.7	1.6	3.0	98.0	1.10

Source: data at the individual firm level gathered by the Colombian National Planning Department and the Ministry of Development, 1969-70.

(a) Does not take account of imported and potentially exportable inputs.

(b) This figure is inflated by the high cost of petrochemical-based raw materials in production of synthetic textiles.

Colombian petrochemical exports are mainly directed to countries within the Latin American Free Trade Area, taking advantage of tariff preferences to offset high cost structures.[9] The poor export performance of Colombian labor-intensive industries mainly reflects the bias against exporting which is introduced by the tariff structure and the overvalued exchange rate.

Clothing, footwear, wooden furniture, textiles, and glass products are in general produced at close to world prices in Colombia, while domestically produced petrochemicals are on average 45 to 50 percent more expensive than imports (Table 2). Overpricing in Colombian petrochemicals would be even greater if local production did not receive two types of implicit subsidy. Producers are able to borrow large amounts of capital from government agencies at subsidized interest rates. And until the early 1970s, the special petroleum exchange rate system ("tasa petrolera") kept petroleum prices artificially low. Preliminary estimates made in 1971 suggested that abolition of the special "tasa petrolera" would raise the price of petrochemicals products by from 5 to 30 percent. When account is also taken of the subsidy via artificially low interest rates, the "true" degree of overpricing in Colombian petrochemicals may be closer to 75 than to 50 percent.[10]

While licensing and prohibition of imports explain how such overpricing can continue, its cause is basically that Colombian petrochemicals plants are on average about one-fifth the size of normal developed country plants producing the same products. Given the importance of economies of scale in the petrochemical industry, the fixed capital costs of producing one ton/year of output in Colombia are therefore on average six times greater than in a typical plant in a developed country (Figure 1). This is not only the case for historical data; it is likely to remain true at least in the near future. The Colombian projects which were in the planning stages in 1971 did not differ significantly in terms of size relative to typical world size from those built previously.[11]

According to *The Economist's* (1970) authoritative survey of the petrochemicals industry, the worldwide trend towards construction of larger and more costly plants is likely to continue at least for some time. Even in the late 1960s, "a fair-sized ethylene cracker . . . cost perhaps U.S. $150 million. Add in an ammonia plant and something of an aromatics complex and one is quickly up to U.S. $500 million."[12] This sum is almost three times as great as Colombian net fixed investment in all branches of industry in 1968, and represents about 40 percent of Colombian gross fixed investment of all kinds in the same year (the most recent one for which data are available).

The best argument in favor of petrochemicals production in

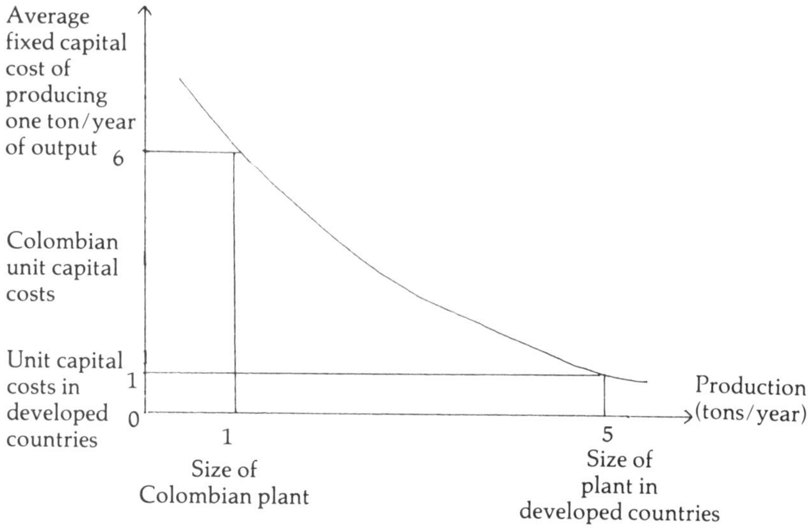

Figure 1—Plant size and production costs in the petrochemical industry in Colombia and developed countries.

Source: Sandoval (1971), *The Economist* (1970), and Nelson (1965)

Colombia is that although it uses huge quantities of one scarce factor, capital, and almost none of the abundant unskilled labor, it does save on the scarce factor of management. Many petrochemicals plants in Colombia are built as simple turn-key operations, and the needs for supervision and coordination of the workers employed are minimal. By contrast, a labor-intensive export-oriented operation using amounts of capital similar to those of a petrochemicals complex would require a great deal of expertise in labor force organization, product design and quality control, and in selling the finished product in export markets. In addition, exports may be confronted with quotas, tariffs, or other barriers in the developed countries, problems which import-substituting petrochemicals producers do not have to face. These are very real problems. Nevertheless, they have been overcome by other less-developed countries which have used exports of labor-intensive products as an engine of growth. There seems little reason to assume that Colombia and the other Andean countries could not do the same.

If the economic case against petrochemicals is so strong, why have Colombia and the other Andean countries placed such great emphasis on petrochemicals production? One important reason seems to be that even within governments, capital is sometimes relatively "sector-specific."

The Colombian government-owned oil company derives a large surplus from its refining and other operations. In the absence of some mechanism for siphoning off this surplus into general government revenues, the company invests it in lines with which its managers and engineers are familiar. Chemical engineers know more about producing petrochemicals than about selling men's shoes in the United States, and are often keen to put their theoretical knowledge into practice.[13] In addition, within governments one often hears of the need to be self-sufficient in "important products," of belief in "modern technology" as the solution to development problems, and of pessimism regarding the possibilities for increasing exports. All of these arguments are used to support production of goods such as petrochemicals.

Second, while there has been some recognition of the need to promote exports, the development policies pursued by the Colombian and other Andean governments have continued on balance to favor import substitution. For example, a Colombian petrochemicals producer whose price is 60 percent above the world price is receiving an implicit subsidy via the tariff and import licensing system of 60 percent. Colombian exporters, on the other hand, receive an export subsidy of only 15-20 percent through the Certificado de Abono Tributario. Producing petrochemicals for the domestic market is therefore made more attractive to private entrepreneurs than exporting labor-intensive goods by the tariff-subsidy structure.

Third, as in many developing countries, Colombian entrepreneurs have easier access and on better terms to fixed capital than to working capital. Thus, industries with a high ratio of fixed to working capital, like petrochemicals, are stimulated more than labor-intensive industries which often have relatively greater needs for working capital.

Finally, as noted above, management and marketing problems are more easily solved in petrochemicals than in labor-intensive exports, and labor problems are likely to be fewer. This last point represents a "natural" advantage of petrochemicals; the first three factors are distortions which could be removed by changes in government policy.

Notes

1. Departamento Nacional de Planeación (1969, Anexo Table II), reports of Andean Group petrochemicals negotiations.

2. 'Sectors' are defined at the old two-digit ISIC level as given in United Nations, *International Standard Industrial Classification of all Economic Activities.* Statistical Papers, Series M. No. 4, Rev. 2, New York, 1968. The old two-digit level roughly corresponds to the new ISIC three-digit level.

3. This section is based on *The Economist* (1970), Nelson (1965), and unpublished data from the Departamento Nacional de Planeación.

4. Departamento Nacional de Planeación (1969, Anexo Table II).

5. These calculations and all others in this paper ignore the fact that part of the fixed capital invested in petrochemicals is provided by foreign funds which would not be available for other purposes, and that working capital needs are greater in labor-intensive industries than in petrochemicals. Taking these points into account (if reliable data were available) would soften the case against petrochemicals a little but would certainly not destroy it. For example, if 50 per cent of the capital invested in petrochemicals is foreign (Table 2) and none of this would have been available for other purposes (an extreme assumption), there would still be at least ten times as many people employed in labor-intensive industries as in petrochemicals for a given domestic fixed investment. On the other hand, the calculations contain some bias in the opposite direction (i.e., toward understating the costs of petrochemicals investments), since the present discounted value of future profit and royalty remittances is not included as a cost, and capital is not shadow-priced.

6. Input-output table for Colombia, 1966 and Morawetz (1971, Table 6). The complementary investments required by indirectly-created jobs are ignored here.

7. Sandoval (1971).

8. Hong Kong, with population one-fourth and G.D.P. one-half of Colombia's, exported $160 million of textiles in 1969. Taiwan, with population and G.D.P. 70 percent of Colombia's, exported $300 million of textiles in the same year.

9. Díaz-Alejandro (1972).

10. On the influence of the "tasa petrolera" see Hutcheson (1971).

11. Sandoval (1971).

12. *The Economist* (1970, p. xii).

13. Engineers often have significant influence in other government agencies as well. For example, in the Colombian National (Economic) Planning Department in 1970, 47 percent of the 160 professionals were engineers by training, 34 percent had studied economics, and the remaining 19 percent were sociologists, lawyers, and others.

References

Departamento Nacional de Planeación, *El Empleo en Colombia: Diagnostico y Recomendaciones de Politica, Tomo II* (DNP-394-URH-Rev. 1, June 1969).

Díaz-Alejandro, Carlos, "Minor Colombian Merchandise Exports," *Yale University Economic Growth Center Discussion Paper No. 149*, July 1972.

Adelman, M.A., and M.B. Zimmerman, "Prices and Profits in Petrochemicals: An Analysis of Investment by Less Developed Countries," *Journal of Industrial Economics*, June 1974.

The Economist, "Petrochemicals: A Survey," Oct. 3, 1970.

Hutcheson, Thomas, *Subsidy to Petroleum Use via the Petroleum Dollar*, unpublished, June 1971.

Morawetz, David, *Concrete Recommendations on Unemployment and Exports of Labor-intensive Manufactured Goods*, unpublished, revised version, May 1971.

Nelson, W.L., "Effect of Plant Size on Cost," *The Oil and Gas Journal*, Jan. 4, 1965.

Sandoval, Fernando, *Desempleo Indirecto en la Industria Textil Causado por los Altos Precios de las Fibras Químicas Producidas en el País*, unpublished, June 1971.

BOOKS WRITTEN UNDER CENTER AUSPICES

The Soviet Bloc, by Zbigniew K. Brzezinski (sponsored jointly with the Russian Research Center), 1960. Harvard University Press. Revised edition, 1967.

The Necessity for Choice, by Henry A. Kissinger, 1961. Harper & Bros.

Rift and Revolt in Hungary, by Ferenc A. Váli, 1961. Harvard University Press.

Strategy and Arms Control, by Thomas C. Schelling and Morton H. Halperin, 1961. Twentieth Century Fund.

United States Manufacturing Investment in Brazil, by Lincoln Gordon and Engelbert L. Grommers, 1962. Harvard Business School.

The Economy of Cyprus, by A. J. Meyer, with Simos Vassiliou (sponsored jointly with the Center for Middle Eastern Studies), 1962. Harvard University Press.

Entrepreneurs of Lebanon, by Yusif A. Sayigh (sponsored jointly with the Center for Middle Eastern Studies), 1962. Harvard University Press.

Communist China 1955-1959: Policy Documents with Analysis, with a foreword by Robert R. Bowie and John K. Fairbank (sponsored jointly with the East Asian Research Center), 1962. Harvard University Press.

Somali Nationalism, by Saadia Touval, 1963. Harvard University Press.

The Dilemma of Mexico's Development, by Raymond Vernon, 1963. Harvard University Press.

Limited War in the Nuclear Age, by Morton H. Halperin, 1963. John Wiley & Sons.

In Search of France, by Stanley Hoffmann *et al.*, 1963. Harvard University Press.

The Arms Debate, by Robert A. Levine, 1963. Harvard University Press.

Africans on the Land, by Montague Yudelman, 1964. Harvard University Press.

Counterinsurgency Warfare, by David Galula, 1964. Frederick A. Praeger, Inc.

People and Policy in the Middle East, by Max Weston Thornburg, 1964. W. W. Norton & Co.

Shaping the Future, by Robert R. Bowie, 1964. Columbia University Press.

Foreign Aid and Foreign Policy, by Edward S. Mason (sponsored jointly with the Council on Foreign Relations), 1964. Harper & Row.

How Nations Negotiate, by Fred Charles Iklé, 1964. Harper & Row.

Public Policy and Private Enterprise in Mexico, edited by Raymond Vernon, 1964. Harvard University Press.

China and the Bomb, by Morton H. Halperin (sponsored jointly with the East Asian Research Center), 1965. Frederick A. Praeger, Inc.

Democracy in Germany, by Fritz Erler (Jodidi Lectures), 1965. Harvard University Press.

The Troubled Partnership, by Henry A. Kissinger (sponsored jointly with the Council on Foreign Relations), 1965. McGraw-Hill Book Co.

The Rise of Nationalism in Central Africa, by Robert I. Rotberg, 1965. Harvard University Press.

Pan-Africanism and East African Integration, by Joseph S. Nye, Jr., 1965. Harvard University Press.

Communist China and Arms Control, by Morton H. Halperin and Dwight H. Perkins (sponsored jointly with the East Asian Research Center), 1965. Frederick A. Praeger, Inc.

Problems of National Strategy, ed. Henry Kissinger, 1965. Frederick A. Praeger, Inc.
Deterrence before Hiroshima: The Airpower Background of Modern Strategy, by George H. Quester, 1966. John Wiley & Sons.
Containing the Arms Race, by Jeremy J. Stone, 1966. M.I.T. Press.
Germany and the Atlantic Alliance: The Interaction of Strategy and Politics, by James L. Richardson, 1966. Harvard University Press.
Arms and Influence, by Thomas C. Schelling, 1966. Yale University Press.
Political Change in a West African State, by Martin Kilson, 1966. Harvard University Press.
Planning without Facts: Lessons in Resource Allocation from Nigeria's Development, by Wolfgang F. Stolper, 1966. Harvard University Press.
Export Instability and Economic Development, by Alasdair I. MacBean, 1966. Harvard University Press.
Foreign Policy and Democratic Politics, by Kenneth N. Waltz (sponsored jointly with the Institute of War and Peace Studies, Columbia University), 1967. Little, Brown & Co.
Contemporary Military Strategy, by Morton H. Halperin, 1967. Little, Brown & Co.
Sino-Soviet Relations and Arms Control, ed. Morton H. Halperin (sponsored jointly with the East Asian Research Center), 1967. M.I.T. Press.
Africa and United States Policy, by Rupert Emerson, 1967. Prentice-Hall.
Elites in Latin America, edited by Seymour M. Lipset and Aldo Solari, 1967. Oxford University Press.
Europe's Postwar Growth, by Charles P. Kindleberger, 1967. Harvard University Press.
The Rise and Decline of the Cold War, by Paul Seabury, 1967. Basic Books.
Student Politics, ed. S.M. Lipset, 1967. Basic Books.
Pakistan's Development: Social Goals and Private Incentives, by Gustav F. Papanek, 1967. Harvard University Press.
Strike a Blow and Die: A Narrative of Race Relations in Colonial Africa, by George Simeon Mwase, ed. Robert I. Rotberg, 1967. Harvard University Press.
Party Systems and Voter Alignments, edited by Seymour M. Lipset and Stein Rokkan, 1967. Free Press.
Agrarian Socialism, by Seymour M. Lipset, revised edition, 1968. Doubleday Anchor.
Aid, Influence, and Foreign Policy, by Joan M. Nelson, 1968. The Macmillan Company.
Development Policy: Theory and Practice, edited by Gustav F. Papanek, 1968. Harvard University Press.
International Regionalism, by Joseph S. Nye, 1968. Little, Brown, & Co.
Revolution and Counterrevolution, by Seymour M. Lipset, 1968. Basic Books.
Political Order in Changing Societies, by Samuel P. Huntington, 1968. Yale University Press.
The TFX Decision: McNamara and the Military, by Robert J. Art, 1968. Little, Brown & Co.
Korea: The Politics of the Vortex, by Gregory Henderson, 1968. Harvard University Press.
Political Development in Latin America, by Martin Needler, 1968. Random House.
The Precarious Republic, by Michael Hudson, 1968. Random House.

The Brazilian Capital Goods Industry, 1929-1964 (sponsored jointly with the Center for Studies in Education and Development), by Nathaniel H. Leff, 1968. Harvard University Press.

Economic Policy-Making and Development in Brazil, 1947-1964, by Nathaniel H. Leff, 1968. John Wiley & Sons.

Turmoil and Transition: Higher Education and Student Politics in India, edited by Philip G. Altbach, 1968. Lalvani Publishing House (Bombay).

German Foreign Policy in Transition, by Karl Kaiser, 1968. Oxford University Press.

Protest and Power in Black Africa, edited by Robert I. Rotberg, 1969. Oxford University Press.

Peace in Europe, by Karl E. Birnbaum, 1969. Oxford University Press.

The Process of Modernization: An Annotated Bibliography on the Sociocultural Aspects of Development, by John Brode, 1969. Harvard University Press.

Students in Revolt, edited by Seymour M. Lipset and Philip G. Altbach, 1969. Houghton Mifflin.

Agricultural Development in India's Districts: The Intensive Agricultural Districts Programme, by Dorris D. Brown, 1970. Harvard University Press.

Authoritarian Politics in Modern Society: The Dynamics of Established One-Party Systems, edited by Samuel P. Huntington and Clement H. Moore, 1970. Basic Books.

Nuclear Diplomacy, by George H. Quester, 1970. Dunellen.

The Logic of Images in International Relations, by Robert Jervis, 1970. Princeton University Press.

Europe's Would-Be Polity, by Leon Lindberg and Stuart A. Scheingold, 1970. Prentice-Hall.

Taxation and Development: Lessons from Colombian Experience, by Richard M. Bird, 1970. Harvard University Press.

Lord and Peasant in Peru: A Paradigm of Political and Social Change, by F. LaMond Tullis, 1970. Harvard University Press.

The Kennedy Round in American Trade Policy: The Twilight of the GATT? by John W. Evans, 1971. Harvard University Press.

Korean Development: The Interplay of Politics and Economics, by David C. Cole and Princeton N. Lyman, 1971. Harvard University Press.

Development Policy II—The Pakistan Experience, edited by Walter P. Falcon and Gustav F. Papanek, 1971. Harvard University Press.

Higher Education in a Transitional Society, by Philip G. Altbach, 1971. Sindhu Publications (Bombay).

Studies in Development Planning, edited by Hollis B. Chenery, 1971. Harvard University Press.

Passion and Politics, by Seymour M. Lipset with Gerald Schaflander, 1971. Little, Brown, & Co.

Political Mobilization of the Venezuelan Peasant, by John D. Powell, 1971. Harvard University Press.

Higher Education in India, edited by Amrik Singh and Philip Altbach, 1971. Oxford University Press (Delhi).

The Myth of the Guerrilla, by J. Bowyer Bell, 1971. Blond (London) and Knopf (New York).

International Norms and War between States: Three Studies in International Politics, by Kjell Goldmann, 1971. Published jointly by Läromedelsförlagen (Sweden) and the Swedish Institute of International Affairs.

Peace in Parts: Integration and Conflict in Regional Organization, by Joseph S. Nye, Jr., 1971. Little, Brown & Co.

Sovereignty at Bay: The Multinational Spread of U.S. Enterprise, by Raymond Vernon, 1971. Basic Books.

Defense Strategy for the Seventies (revision of *Contemporary Military Strategy*), by Morton H. Halperin, 1971. Little, Brown & Co.

Peasants Against Politics: Rural Organization in Brittany, 1911-1967, by Suzanne Berger, 1972. Harvard University Press.

Transnational Relations and World Politics, edited by Robert O. Keohane and Joseph S. Nye, Jr., 1972. Harvard University Press.

Latin American University Students: A Six Nation Study, by Arthur Liebman, Kenneth N. Walker, and Myron Glazer, 1972. Harvard University Press.

The Politics of Land Reform in Chile, 1950-1970: Public Policy, Political Institutions, and Social Change, by Robert R. Kaufman, 1972. Harvard University Press.

The Boundary Politics of Independent Africa, by Saadia Touval, 1972. Harvard University Press.

The Politics of Nonviolent Action, by Gene E. Sharp, 1973. Porter Sargent.

System 37 Viggen: Arms, Technology, and the Domestication of Glory, by Ingemar Dörfer, 1973. Universitetsforlaget (Oslo).

University Students and African Politics, by William John Hanna. 1974. Africana Publishing Company.

Organizing the Transnational: The Experience with Transnational Enterprise in Advanced Technology, by M. S. Hochmuth, 1974. Sijthoff (Leiden).

Becoming Modern, by Alex Inkeles and David H. Smith, 1974. Harvard University Press.

Multinational Corporations and the Politics of International Dependence: Copper in Chile, 1945-1973, by Theodore Moran, 1974. Princeton University Press.

The Andean Group: A Case Study in Economic Integration among Developing Countries, by David Morawetz, 1974. M.I.T. Press.

Kenya: The Politics of Participation and Control, by Henry Bienen, 1974. Princeton University Press.

Land Reform and Politics: A Comparative Analysis, by Hung-chao Tai, 1974. University of California Press.

Big Business and the State: Changing Relations in Western Europe, edited by Raymond Vernon, 1974. Harvard University Press.

Harvard Studies in International Affairs*

[*formerly Occasional Papers in International Affairs*]

† 1. *A Plan for Planning: The Need for a Better Method of Assisting Underdeveloped Countries on Their Economic Policies,* by Gustav F. Papanek, 1961.
† 2. *The Flow of Resources from Rich to Poor,* by Alan D. Neale, 1961.
† 3. *Limited War: An Essay on the Development of the Theory and an Annotated Bibliography,* by Morton H. Halperin, 1962.
† 4. *Reflections on the Failure of the First West Indian Federation,* by Hugh W. Springer, 1962.
5. *On the Interaction of Opposing Forces under Possible Arms Agreements,* by Glenn A. Kent, 1963. 36 pp. $1.25.
† 6. *Europe's Northern Cap and the Soviet Union,* by Nils Örvik, 1963.
7. *Civil Administration in the Punjab: An Analysis of a State Government in India,* by E. N. Mangat Rai, 1963. 82 pp. $1.75.
8. *On the Appropriate Size of a Development Program,* by Edward S. Mason, 1964. 24 pp. $1.00.
9. *Self-Determination Revisited in the Era of Decolonization,* by Rupert Emerson, 1964. 64 pp. $1.75.
10. *The Planning and Execution of Economic Development in Southeast Asia,* by Clair Wilcox, 1965. 37 pp. $1.25.
11. *Pan-Africanism in Action,* by Albert Tevoedjre, 1965. 88 pp. $2.50.
12. *Is China Turning In?,* by Morton Halperin, 1965. 34 pp. $1.25.
†13. *Economic Development in India and Pakistan,* by Edward S. Mason, 1966.
14. *The Role of the Military in Recent Turkish Politics,* by Ergun Özbudun, 1966. 54 pp. $1.75.
†15. *Economic Development and Individual Change: A Social-Psychological Study of the Comilla Experiment in Pakistan,* by Howard Schuman, 1967.
16. *A Select Bibliography on Students, Politics, and Higher Education,* by Philip G. Altbach, UMHE Revised Edition, 1970. 65 pp. $2.75.
17. *Europe's Political Puzzle: A Study of the Fouchet Negotiations and the 1963 Veto,* by Alessandro Silj, 1967. 178 pp. $3.50.
18. *The Cap and the Straits: Problems of Nordic Security,* by Jan Klenberg, 1968. 19 pp. $1.25.
19. *Cyprus: The Law and Politics of Civil Strife,* by Linda B. Miller, 1968. 97 pp. $3.00.
†20. *East and West Pakistan: A Problem in the Political Economy of Regional Planning,* by Md. Anisur Rahman, 1968.
†21. *Internal War and International Systems: Perspectives on Method,* by George A. Kelley and Linda B. Miller, 1969.

*Available from Harvard University Center for International Affairs, 6 Divinity Avenue, Cambridge, Massachusetts 02138
†Out of print. May be ordered from AMS Press, Inc., 56 East 13th Street, New York, N.Y. 10003

†22. *Migrants, Urban Poverty, and Instability in Developing Nations,* by Joan M. Nelson, 1969. 81 pp.
23. *Growth and Development in Pakistan, 1955-1969,* by Joseph J. Stern and Walter P. Falcon, 1970. 94 pp. $3.00.
24. *Higher Education in Developing Countries: A Select Bibliography,* by Philip G. Altbach, 1970. 118 pp. $4.00.
25. *Anatomy of Political Institutionalization: The Case of Israel and Some Comparative Analyses,* by Amos Perlmutter, 1970. 60 pp. $2.50.
†26. *The German Democratic Republic from the Sixties to the Seventies,* by Peter Christian Ludz, 1970. 100 pp.
27. *The Law in Political Integration: The Evolution and Integrative Implications of Regional Legal Processes in the European Community,* by Stuart A. Scheingold, 1971. 63 pp. $2.50.
28. *Psychological Dimensions of U.S.-Japanese Relations,* by Hiroshi Kitamura, 1971. 46 pp. $2.00.
29. *Conflict Regulation in Divided Societies,* by Eric A. Nordlinger, 1972. 137 pp. $4.25.
30. *Israel's Political-Military Doctrine,* by Michael I. Handel, 1973. 101 pp. $3.25.
31. *Italy, NATO, and the European Community: The Interplay of Foreign Policy and Domestic Politics,* by Primo Vanicelli, 1974. 67 + x pp. $3.25.
32. *The Choice of Technology in Developing Countries: Some Cautionary Tales,* by C. Peter Timmer, John W. Thomas, Louis T. Wells, Jr., and David Morawetz, 1975. 114 pp. $3.45.
33. *The International Role of the Communist Parties of Italy and France,* by Donald L. M. Blackmer and Annie Kriegel, 1975. 66 + x pp. $2.75.
34. *The Hazards of Peace: A European View of Detente,* by Juan Cassiers, 1975.